AF307251

Marx Joyce
Abbott Hardy Machiavelli Cheston Cooper Austen
Defoe Melville Montaigne Haggard Emerson Hugo
Stoker Carroll Christie Byron Eliot Grimm
Wilde Maupassant Engels Schiller Molière
Garnett Fitzgerald Hawthorne Smith Kafka
Goethe Einstein Dostoyevsky Hall
Baum Cotton Kipling Doyle Willis
Henry Nietzsche
Leslie Dumas Flaubert Turgenev Balzac
Stockton Vatsyayana Crane
Burroughs Verne
Curtis Tocqueville Gogol Vinci
Homer Widger Tolstoy Whitman Busch
Darwin Thoreau Twain
Potter Freud Zola Scott Plato Harte
Kant Jowett Stevenson Dickens Burton Hesse
Andersen Cooke Cervantes
London Descartes Voltaire
Poe Aristotle Wells Cooke
Hale James Hastings Irving
Bunner Shakespeare
Richter Chekhov Chambers
Doré Dante Shaw Benedict Alcott
Swift Pushkin
Newton Wodehouse

tredition

tredition was established in 2006 by Sandra Latusseck and Soenke Schulz. Based in Hamburg, Germany, tredition offers publishing solutions to authors and publishing houses, combined with world-wide distribution of printed and digital book content. tredition is uniquely positioned to enable authors and publishing houses to create books on their own terms and without conventional manu-facturing risks.

For more information please visit: www.tredition.com

TREDITION CLASSICS

This book is part of the TREDITION CLASSICS series. The creators of this series are united by passion for literature and driven by the intention of making all public domain books available in printed format again - worldwide. Most TREDITION CLASSICS titles have been out of print and off the bookstore shelves for decades. At tredition we believe that a great book never goes out of style and that its value is eternal. Several mostly non-profit literature projects provide content to tredition. To support their good work, tredition donates a portion of the proceeds from each sold copy. As a reader of a TREDITION CLASSICS book, you support our mission to save many of the amazing works of world literature from oblivion. See all available books at www.tredition.com.

 Project Gutenberg

The content for this book has been graciously provided by Project Gutenberg. Project Gutenberg is a non-profit organization founded by Michael Hart in 1971 at the University of Illinois. The mission of Project Gutenberg is simple: To encourage the creation and distribution of eBooks. Project Gutenberg is the first and largest collection of public domain eBooks.

The Dominion of the Air, the story of aerial navigation

John Mackenzie Bacon

Imprint

This book is part of TREDITION CLASSICS

Author: John Mackenzie Bacon
Cover design: Buchgut, Berlin – Germany

Publisher: tredition GmbH, Hamburg - Germany
ISBN: 978-3-8424-3889-7

www.tredition.com
www.tredition.de

THE DOMINION OF THE AIR

The Story of Aerial Navigation

by J. M. Bacon

Contents

CHAPTER I.   THE DAWN OF AERONAUTICS.

CHAPTER II.   THE INVENTION OF THE BALLOON.

CHAPTER III.   THE FIRST BALLOON ASCENT IN ENGLAND.

CHAPTER IV.   THE DEVELOPMENT OF BALLOON PHILOSOPHY.

CHAPTER V.   SOME FAMOUS EARLY VOYAGERS.

CHAPTER VI.   CHARLES GREEN AND THE NASSAU BALLOON.

CHAPTER VII. CHARLES GREEN — FURTHER ADVENTURES.

CHAPTER VIII. JOHN WISE — THE AMERICAN AERONAUT.

CHAPTER IX. EARLY METHODS AND IDEAS.

CHAPTER X.   THE COMMENCEMENT OF A NEW ERA.

CHAPTER XI. THE BALLOON IN THE SERVICE OF SCIENCE.

CHAPTER XII. HENRY COXWELL AND HIS CONTEMPORARIES.

CHAPTER XIII. SOME NOTEWORTHY ASCENTS.

CHAPTER XIV. THE HIGHEST ASCENT ON RECORD.

CHAPTER XV. FURTHER SCIENTIFIC VOYAGES OF GLAISHER AND COXWELL.

CHAPTER XVI. SOME FAMOUS FRENCH AERONAUTS.

CHAPTER XVII.	ADVENTURE AND ENTERPRISE.
CHAPTER XVIII.	THE BALLOON IN THE SIEGE OF PARIS.
CHAPTER XIX.	THE TRAGEDY OF THE ZENITH — THE NAVIGABLE BALLOON
CHAPTER XX.	A CHAPTER OF ACCIDENTS.
CHAPTER XXI.	THE COMING OF THE FLYING MACHINE.
CHAPTER XXII.	THE STORY OF THE SPENCERS.
CHAPTER XXIII.	NEW DEPARTURES IN AEROSTATION.
CHAPTER XXIV.	ANDREE AND HIS VOYAGES
CHAPTER XXV.	THE MODERN AIRSHIP — IN SEARCH OF THE LEONIDS.
CHAPTER XXVI.	RECENT AERONAUTICAL EVENTS.
CHAPTER XXVII.	THE POSSIBILITIES OF BALLOONS IN WARFARE.
CHAPTER XXVIII.	THE CONSTITUTION OF THE AIR.
CHAPTER XXIX.	CONCLUSION.

CHAPTER I. THE DAWN OF AERONAUTICS.

"He that would learn to fly must be brought up to the constant practice of it from his youth, trying first only to use his wings as a tame goose will do, so by degrees learning to rise higher till he attain unto skill and confidence."

So wrote Wilkins, Bishop of Chester, who was reckoned a man of genius and learning in the days of the Commonwealth. But so soon as we come to inquire into the matter we find that this good Bishop was borrowing from the ideas of others who had gone before him; and, look back as far as we will, mankind is discovered to have entertained persistent and often plausible ideas of human flight. And those ideas had in some sort of way, for good or ill, taken practical shape. Thus, as long ago as the days when Xenophon was leading back his warriors to the shores of the Black Sea, and ere the Gauls had first burned Rome, there was a philosopher, Archytas, who invented a pigeon which could fly, partly by means of mechanism, and partly also, it is said, by aid of an aura or spirit. And here arises a question. Was this aura a gas, or did men use it as spiritualists do today, as merely a word to conjure with?

Four centuries later, in the days of Nero, there was a man in Rome who flew so well and high as to lose his life thereby. Here, at any rate, was an honest man, or the story would not have ended thus; but of the rest—and there are many who in early ages aspired to the attainment of flight—we have no more reason to credit their claims than those of charlatans who flourish in every age.

In medieval times we are seriously told by a saintly writer (St. Remigius) of folks who created clouds which rose to heaven by means of "an earthen pot in which a little imp had been enclosed." We need no more. That was an age of flying saints, as also of flying dragons. Flying in those days of yore may have been real enough to the multitude, but it was at best delusion. In the good old times it did not need the genius of a Maskelyne to do a "levitation" trick. We can picture the scene at a "flying seance." On the one side the decidedly professional showman possessed of sufficient low cunning; on the other the ignorant and highly superstitious audience, eager to hear or see some new thing—the same audience that, deceived by a simple trick of schoolboy science, would listen to supernatural voices in their groves, or oracular utterances in their temples, or watch the urns of Bacchus fill themselves with wine. Surely for their eyes it would need no more than the simplest phantasmagoria, or maybe only a little black thread, to make a pigeon rise and fly.

It is interesting to note, however, that in the case last cited there is unquestionably an allusion to some crude form of firework, and what more likely or better calculated to impress the ignorant! Our firework makers still manufacture a "little Devil." Pyrotechnic is as old as history itself; we have an excellent description of a rocket in a document at least as ancient as the ninth century. And that a species of pyrotechny was resorted to by those who sought to imitate flight we have proof in the following recipe for a flying body given by a Doctor, eke a Friar, in Paris in the days of our King John: —

"Take one pound of sulphur, two pounds of willowcarbon, six pounds of rock salt ground very fine in a marble mortar. Place, when you please, in a covering made of flying papyrus to produce thunder. The covering in order to ascend and float away should be long, graceful, well filled with this fine powder; but to produce thunder the covering should be short, thick, and half full."

Nor does this recipe stand alone. Take another sample, of which chapter and verse are to be found in the MSS. of a Jesuit, Gaspard Schott, of Palermo and Rome, born three hundred years ago: —

"The shells of hen-eggs, if properly filled and well secured against the penetration of the air, and exposed to solar rays, will ascend to the skies and sometimes suffer a natural change. And if the eggs of the larger description of swans, or leather balls stitched with fine thongs, be filled with nitre, the purest sulphur quicksilver, or kindred materials which rarify by their caloric energy, and if they externally resemble pigeons, they will easily be mistaken for flying animals."

Thus it would seem that, hunting back in history, there were three main ideas on which would-be aeronauts of old exercised their ingenuity. There was the last-mentioned method, which, by the way, Jules Verne partly relies on when he takes his heroes to the moon, and which in its highest practical development may be seen annually on the night of "Brock's Benefit" at the Crystal Palace. There is, again, the "tame goose" method, to which we must return presently; and, lastly, there is a third method, to which, as also to the brilliant genius who conceived it, we must without further delay be introduced. This may be called the method of "a hollow globe."

Roger Bacon, Melchisedeck-fashion, came into existence at Ilchester in 1214 of parentage that is hard to trace. He was, however, a born philosopher, and possessed of intellect and penetration that placed him incalculably ahead of his generation. A man of marvellous insight and research, he grasped, and as far as possible carried out, ideas which dawned on other men only after centuries. Thus, many of his utterances have been prophetic. It is probable that among his chemical discoveries he re-invented gunpowder. It is certain that he divined the properties of a lens, and diving deep into experimental and mechanical sciences, actually foresaw the time when, in his own words, "men would construct engines to traverse land and water with great speed and carry with them persons and merchandise." Clearly in his dreams Bacon saw the Atlantic not merely explored, but on its bosom the White Star liners breaking records, contemptuous of its angriest seas. He saw, too, a future Dumont circling in the air, and not only in a dead calm, but holding his own with the feathered race. He tells his dream thus: "There may be made some flying instrument so that a man sitting in the middle of the instrument and turning some mechanism may put in motion some artificial wings which may beat the air like a bird flying."

But he lived too long before his time. His ruin lay not only in his superior genius, but also in his fearless outspokenness. He presently fell under the ban of the Church, through which he lost alike his liberty and the means of pursuing investigation. Had it been otherwise we may fairly believe that the "admirable Doctor," as he was called, would have been the first to show mankind how to navigate the air. His ideas are perfectly easy to grasp. He conceived that the air was a true fluid, and as such must have an upper limit, and it would be on this upper surface, he supposed, as on the bosom of the ocean, that man would sail his air-ship. A fine, bold guess truly. He would watch the cirrus clouds sailing grandly ten miles above him on some stream that never approached nearer. Up there, in his imagination, would be tossing the waves of our ocean of air. Wait for some little better cylinders of oxygen and an improved foot-warmer, and a future Coxwell will go aloft and see; but as to an upper sea, it is truly there, and we may visit and view its sun-lit tossing billows stretching out to a limitless horizon at such times as the nether world is shrouded in densest gloom. Bacon's method of reaching such an upper sea as he postulated was, as we have said, by a hollow globe.

"The machine must be a large hollow globe, of copper or other suitable metal, wrought extremely thin so as to have it as light as possible," and "it must be filled with ethereal air or liquid fire." This was written in the thirteenth century, and it is scarcely edifying to find four hundred years after this the Jesuit Father Lana, who contrived to make his name live in history as a theoriser in aeronautics, arrogating to himself the bold conception of the English Friar, with certain unfortunate differences, however, which in fairness we must here clearly point out. Lana proclaimed his speculations standing on a giant's shoulders. Torricelli, with his closed bent tube, had just shown the world how heavily the air lies above us. It then required little mathematical skill to calculate what would be the lifting power of any vessel void of air on the earth's surface. Thus Lana proposed the construction of an air ship which possibly because of its picturesquesness has won him notoriety. But it was a fraud. We have but to conceive a dainty boat in which the aeronaut would sit at ease handling a little rudder and a simple sail. These, though a schoolboy would have known better, he thought would guide his vessel when in the air.

So much has been claimed for Father Lana and his mathematical and other attainments that it seems only right to insist on the weakness of his reasoning. An air ship simply drifting with the wind is incapable of altering its course in the slightest degree by either sail or rudder. It is simply like a log borne along in a torrent; but to compare such a log properly with the air ship we must conceive it WHOLLY submerged in the water and having no sail or other appendage projecting into the air, which would, of course, introduce other conditions. If, however, a man were to sit astride of the log and begin to propel it so that it travels either faster or slower than the stream, then in that case, either by paddle or rudder, the log could be guided, and the same might be said of Lana's air boat if only he had thought of some adequate paddle, fan, or other propeller. But he did not. One further explanatory sentence may here be needed; for we hear of balloons which are capable of being guided to a small extent by sail and rudder. In these cases, however, the rudder is a guide rope trailing on earth or sea, so introducing a fresh element and fresh conditions which are easy to explain.

Suppose a free balloon drifting down the wind to have a sail suddenly hoisted on one side, what happens? The balloon will simply swing till this sail is in front, and thus continue its straightforward course. Suppose, however, that as soon as the side sail is hoisted a trail rope is also dropped aft from a spar in the rigging. The tendency of the sail to fly round in front is now checked by the dragging rope, and it is constrained to remain slanting at an angle on one side; at the same time the rate of the balloon is reduced by the dragging rope, so that it travels slower than the wind, which, now acting on its slant sail, imparts a certain sidelong motion much as it does in the case of a sailing boat.

Lana having in imagination built his ship, proceeds to make it float up into space, for which purpose he proposes four thin copper globes exhausted of air. Had this last been his own idea we might have pardoned him. We have, however, pointed out that it was not, and we must further point out that in copying his great predecessor he fails to see that he would lose enormous advantage by using four globes instead of one. But, beyond all, he failed to see what the master genius of Bacon saw clearly—that his thin globes when exhausted must infallibly collapse by virtue of that very pressure of the air which he sought to make use of.

It cannot be too strongly insisted on that if the too much belauded speculations of Lana have any value at all it is that they throw into stronger contrast the wonderful insight of the philosopher who so long preceded him. By sheer genius Bacon had foreseen that the emptied globe must be filled with SOMETHING, and for this something he suggests "ethereal air" or "liquid fire," neither of which, we contend, were empty terms. With Bacon's knowledge of experimental chemistry it is a question, and a most interesting one, whether he had not in his mind those two actual principles respectively of gas and air rarefied by heat on which we launch our balloons into space today.

Early progress in any art or science is commonly intermittent. It was so in the story of aeronautics. Advance was like that of the incoming tide, throwing an occasional wave far in front of its rising flood. It was a phenomenal wave that bore Roger Bacon and left his mark on the sand where none other approached for centuries. In those centuries men were either too priest-ridden to lend an ear to Science, or, like children, followed only the Will-o'-the-Wisp floating above the quagmire which held them fast. They ran after the stone that was to turn all to gold, or the elixir that should conquer death, or the signs in the heavens that should foretell their destinies; and the taint of this may be traced even when the dark period that followed was clearing away. Four hundred years after Roger's death, his illustrious namesake, Francis Bacon, was formulating his Inductive Philosophy, and with complete cock-sureness was teaching mankind all about everything. Let us look at some of his utterances which may help to throw light on the way he regarded the problem we are dealing with.

"It is reported," Francis Bacon writes, "that the Leucacians in ancient time did use to precipitate a man from a high cliffe into the sea; tying about him, with strings, at some distance, many great fowles; and fixing unto his body divers feathers, spread, to breake the fall. Certainly many birds of good wing (as Kites and the like) would beare up a good weight as they flie. And spreading of feathers, thin and close, and in great breadth, will likewise beare up a great weight, being even laid without tilting upon the sides. The further extension of this experiment of flying may be thought upon."

To say the least, this is hardly mechanical. But let us next follow the philosopher into the domain of Physics. Referring to a strange assertion, that "salt water will dissolve salt put into it in less time than fresh water will dissolve it," he is at once ready with an explanation to fit the case. "The salt," he says, "in the precedent water doth by similitude of substance draw the salt new put in unto it." Again, in his finding, well water is warmer in winter than summer, and "the cause is the subterranean heat which shut close in (as in winter) is the more, but if it perspire (as it doth in summer) it is the less." This was Bacon the Lord. What a falling off—from the experimentalist's point of view—from Bacon the Friar! We can fancy him watching a falcon poised motionless in the sky, and reflecting on that problem which to this day fairly puzzles our ablest scientists, settling the matter in a sentence: "The cause is that feathers doe possess upward attractions." During four hundred years preceding Lord Verulam philosophers would have flown by aid of a broomstick. Bacon himself would have merely parried the problem with a platitude!

At any rate, physicists, even in the brilliant seventeenth century, made no material progress towards the navigation of the air, and thus presently let the simple mechanic step in before them. Ere that century had closed something in the nature of flight had been accomplished. It is exceedingly hard to arrive at actual fact, but it seems pretty clear that more than one individual, by starting from some eminence, could let himself fall into space and waft himself away for some distance with fair success and safety, It is stated that an English Monk, Elmerus, flew the space of a furlong from a tower in Spain, a feat of the same kind having been accomplished by another adventurer from the top of St. Mark's at Venice.

In these attempts it would seem that the principle of the parachute was to some extent at least brought into play. If also circumstantial accounts can be credited, it would appear that a working model of a flying machine was publicly exhibited by one John Muller before the Emperor Charles V. at Nuremberg. Whatever exaggeration or embellishment history may be guilty of it is pretty clear that some genuine attempts of a practical and not unsuccessful nature had been made here and there, and these prompted the flowery and visionary Bishop Wilkins already quoted to predict confidently that the day was approaching when it "would be as common for a man to call for his wings as for boots and spurs."

We have now to return to the "tame goose" method, which found its best and boldest exponent in a humble craftsman, by name Besnier, living at Sable, about the year 1678. This mechanical genius was by trade a locksmith, and must have been possessed of sufficient skill to construct an efficient apparatus out of such materials as came to his hand, of the simplest possible design. It may be compared to the earliest type of bicycle, the ancient "bone shaker," now almost forgotten save by those who, like the writer, had experience of it on its first appearance. Besnier's wings, as it would appear, were essentially a pair of double-bladed paddles and nothing more, roughly resembling the double-paddle of an old-fashioned canoe, only the blades were large, roughly rectangular, and curved or hollowed. The operator would commence by standing erect and balancing these paddles, one on each shoulder, so that the hollows of the blades should be towards the ground. The forward part of each paddle was then grasped by the hands, while the hinder part of each was connected to the corresponding leg. This, presumably, would be effected after the arms had been raised vertically, the leg attachment being contrived in some way which experience would dictate.

The flyer was now fully equipped, and nothing remained for him save to mount some eminence and, throwing himself forward into space and assuming the position of a flying bird, to commence flapping and beating the air with a reciprocal motion. First, he would buffet the air downwards with the left arm and right leg simultaneously, and while these recovered their position would strike with the right hand and left leg, and so on alternately. With this crude method the enterprising inventor succeeded in raising himself by short stages from one height to another, reaching thus the top of a house, whence he could pass over others, or cross a river or the like.

The perfecting of his system became then simply a question of practice and experience, and had young athletes only been trained from early years to the new art it seems reasonable to suppose that some crude approach to human flight would have been effected. Modifications and improvements in construction would soon have suggested themselves, as was the case with the bicycle, which in its latest developments can scarcely be recognised as springing from the primitive "bone-shaker" of thirty-three years ago. We would suggest the idea to the modern inventor. He will in these days, of course, find lighter materials to hand. Then he will adopt some link motion for the legs in place of leather thongs, and will hinge the paddle blades so that they open out with the forward stroke, but collapse with the return. Then look on another thirty-three years—a fresh generation—and our youth of both sexes may find a popular recreation in graceful aerial exercise. The pace is not likely to be excessive, and molestations from disguised policemen—not physically adapted, by the way, to rapid flight—need not be apprehended.

One of the best tests of Besnier's measure of success is supplied by the fact that he had pupils as well as imitators. First on this list must be mentioned a Mr. Baldwin, a name which, curiously enough, twice over in modern times comes into the records of bold aerial exploits. This individual, it appears, purchased a flying outfit of Besnier himself, and surpassed his master in achievement. A little later one Dante contrived some modification of the same apparatus, with which he pursued the new mode of progress till he met with a fractured thigh.

But whatever the imitators of Besnier may have accomplished, to the honest smith must be accorded the full credit of their success, and with his simple, but brilliant, record left at flood mark, the tide of progress ebbed back again, while mankind ruminated over the great problem in apparent inactivity. But not for long. The air-pump about this period was given to the world, and chemists were already busy investigating the nature of gases. Cavallo was experimenting on kindred lines, while in our own land the rival geniuses of Priestley and Cavendish were clearing the way to make with respect to the atmosphere the most important discovery yet dreamed of. In recording this dawn of a new era, however, we should certainly not forget how, across the Atlantic, had arisen a Rumford and a Franklin, whose labours were destined to throw an all-important sidelight on the pages of progress which we have now to chronicle.

CHAPTER II. THE INVENTION OF THE BALLOON.

It was a November night of the year 1782, in the little town of Annonay, near Lyons. Two young men, Stephen and Joseph Montgolfier, the representatives of a firm of paper makers, were sitting together over their parlour fire. While watching the smoke curling up the chimney one propounded an idea by way of a sudden inspiration: "Why shouldn't smoke be made to raise bodies into the air?"

The world was waiting for this utterance, which, it would seem, was on the tip of the tongue with many others. Cavendish had already discovered what he designated "inflammable air," though no one had as yet given it its later title of hydrogen gas. Moreover, in treating of this gas—Dr. Black of Edinburgh, as much as fifteen years before the date we have now arrived at, had suggested that it should be made capable of raising a thin bladder in the air. With a shade more of good fortune, or maybe with a modicum more of leisure, the learned Doctor would have won the invention of the balloon for his own country. Cavallo came almost nearer, and actually putting the same idea into practice, had succeeded in the spring of 1782 in making soap bubbles blown with hydrogen gas float upwards. But he had accomplished no more when, as related, in the autumn of the same year the brothers Montgolfier conceived the notion of making bodies "levitate" by the simpler expedient of filling them with smoke.

This was the crude idea, the application of which in their hands was soon marked with notable success. Their own trade supplied ready and suitable materials for a first experiment, and, making an oblong bag of thin paper a few feet in length, they proceeded to introduce a cloud of smoke into it by holding crumpled paper kindled in a chafing dish beneath the open mouth. What a subject is there here for an imaginative painter! As the smoky cloud formed within, the bag distended itself, became buoyant, and presently floated to the ceiling. The simple trial proved a complete success, due, as it appeared to them, to the ascensive power of a cloud of smoke.

An interesting and more detailed version of the story is extant. While the experiment was in progress a neighbour, the widow of a tradesman who had been connected in business with the firm, seeing smoke escaping into the room, entered and stood watching the proceedings, which were not unattended with difficulties. The bag, half inflated, was not easy to hold in position over the chafing dish, and rapidly cooled and collapsed on being removed from it. The widow noting this, as also the perplexity of the young men, suggested that they should try the result of tying the dish on at the bottom of the bag. This was the one thing wanted to secure success, and that good lady, whose very name is unhappily lost, deserves an honoured place in history. It was unquestionably the adoption of her idea which launched the first balloon into space.

The same experiment repeated in the open air proving a yet more pronounced success, more elaborate trials were quickly developed, and the infant balloon grew fast. One worthy of the name, spherical in shape and of some 600 cubic feet capacity, was now made and treated as before, with the result that ere it was fully inflated it broke the strings that held it and sailed away hundreds of feet into the air. The infant was fast becoming a prodigy. Encouraged by their fresh success, the inventors at once set about preparations for the construction of a much larger balloon some thirty-five feet diameter (that is, of about 23,000 cubic feet capacity), to be made of linen lined with paper and this machine, launched on a favourable day in the following spring, rose with great swiftness to fully a thousand feet, and travelled nearly a mile from its starting ground.

Enough; the time was already ripe for a public demonstration of the new invention, and accordingly the 5th of the following June witnessed the ascent of the same balloon with due ceremony and advertisement. Special pains were taken with the inflation, which was conducted over a pit above which the balloon envelope was slung; and in accordance with the view that smoke was the chief lifting power, the fuel was composed of straw largely mixed with wool. It is recorded that the management of the furnace needed the attention of two men only, while eight men could hardly hold the impatient balloon in restraint. The inflation, in spite of the fact that the fuel chosen was scarcely the best for the purpose, was conducted remarkable expedition, and on being released, the craft travelled one and a half miles into the air, attaining a height estimated at over 6,000 feet.

From this time the tide of events in the aeronautical world rolls on in full flood, almost every half-year marking a fresh epoch, until a new departure in the infant art of ballooning was already on the point of being reached. It had been erroneously supposed that the ascent of the Montgolfier balloon had been due, not to the rarefaction of the air within it—which was its true cause—but to the evolution of some light gas disengaged by the nature of the fuel used. It followed, therefore, almost as a matter of course, that chemists, who, as stated in the last chapter, were already acquainted with so-called "inflammable air," or hydrogen gas, grasped the fact that this gas would serve better than any other for the purposes of a balloon. And no sooner had the news of the Montgolfiers' success reached Paris than a subscription was raised, and M. Charles, Professor of Experimental Philosophy, was appointed, with the assistance of M. Roberts, to superintend the construction of a suitable balloon and its inflation by the proposed new method.

The task was one of considerable difficulty, owing partly to the necessity of procuring some material which would prevent the escape of the lightest and most subtle gas known, and no less by reason of the difficulty of preparing under pressure a sufficient quantity of gas itself. The experiment, sound enough in theory, was eventually carried through after several instructive failures. A suitable material was found in "lustring," a glossy silk cloth varnished with a solution of caoutchouc, and this being formed into a balloon only thirteen feet in diameter and fitted without other aperture than a stopcock, was after several attempts filled with hydrogen gas prepared in the usual way by the action of dilute sulphuric acid on scrap iron.

The preparations completed, one last and all-important mistake was made by closing the stop-cock before the balloon was dismissed, the disastrous and unavoidable result of this being at the time overlooked.

On August 25, 1783, the balloon was liberated on the Champ de Mars before an enormous concourse, and in less than two minutes had reached an elevation of half a mile, when it was temporarily lost in cloud, through which, however, it penetrated, climbing into yet higher cloud, when, disappearing from sight, it presently burst and descended to earth after remaining in the air some three-quarters of an hour.

The bursting of this little craft taught the future balloonist his first great lesson, namely, that on leaving earth he must open the neck of his balloon; and the reason of this is obvious. While yet on earth the imprisoned gas of a properly filled balloon distends the silk by virtue of its expansive force, and in spite of the enormous outside pressure which the weight of air exerts upon it. Then, as the balloon rises high in the air and the outside pressure grows less, the struggling gas within, if allowed no vent, stretches the balloon more and more until the slender fabric bursts under the strain.

At the risk of being tedious, we have dwelt at some length on the initial experiments which in less than a single year had led to the discovery and development of two distinct methods—still employed and in competition with each other—of dismissing balloons into the heavens. We are now prepared to enter fully into the romantic history of our subject which from this point rapidly unfolds itself.

Some eleven months only after the two Montgolfiers were discovered toying with their inflated paper bag, the younger of the two brothers was engaged to make an exhibition of his new art before the King at Versailles, and this was destined to be the first occasion when a balloon was to carry a living freight into the sky. The stately structure, which was gorgeously decorated, towered some seventy feet into the air, and was furnished with a wicker car in which the passengers were duly installed. These were three in number, a sheep, a cock, and a duck, and amid the acclamations of the multitude, rose a few hundred feet and descended half a mile away. The cock was found to have sustained an unexplained mishap: its leg was broken; but the sheep was feeding complacently, and the duck was quacking with much apparent satisfaction.

Now, who among mortals will come forward and win the honour of being the first to sail the skies? M. Pilitre de Rozier at once volunteered, and by the month of November a new air ship was built, 74 feet high, 48 feet in largest diameter, and 15 feet across the neck, outside which a wicker gallery was constructed, while an iron brazier was slung below all. But to trim the boat properly two passengers were needed, and de Rozier found a ready colleague in the Marquis d'Arlandes. By way of precaution, de Rozier made a few preliminary ascents with the balloon held captive, and then the two intrepid Frenchmen took their stand on opposite sides of the gallery, each furnished with bundles of fuel to feed the furnace, each also carrying a large wet sponge with which to extinguish the flames whenever the machine might catch fire. On casting off the balloon rose readily, and reaching 3,000 feet, drifted away on an upper current.

The rest of the narrative, much condensed from a letter of the Marquis, written a week later, runs somewhat thus: "Our departure was at fifty-four minutes past one, and occasioned little stir among the spectators. Thinking they might be frightened and stand in need of encouragement, I waved my arm. M. de Rozier cried, 'You are doing nothing, and we are not rising!' I stirred the fire, and then began to scan the river, but Pilitre cried again, 'See the river; we are dropping into it!' We again urged the fire, but still clung to the river bed. Presently I heard a noise in the upper part of the balloon, which gave a shock as though it had burst. I called to my companion, 'Are you dancing?' The balloon by now had many holes burned in it, and using my sponge I cried that we must descend. My companion, however, explained that we were over Paris, and must now cross it. Therefore, raising the fire once more, we turned south till we passed the Luxemburg, when, extinguishing the flame, the balloon came down spent and empty."

Daring as was this ascent, it was in achievement eclipsed two months later at Lyons, when a mammoth balloon, 130 feet in height and lifting 18 tons, was inflated in seventeen minutes, and ascended with no less than seven passengers. When more than half a mile aloft this machine, which was made of too slender material for its huge size, suddenly developed a rent of half its length, causing it to descend with immense velocity; but without the smallest injury to any of the passengers. This was a memorable performance, and the account, sensational as it may read, is by no means unworthy of credit; for, as will be seen hereafter, a balloon even when burst or badly torn in midair may, on the principle of the parachute, effect its own salvation.

In the meanwhile, the rival balloon of hydrogen gas—the Charliere, as it has been called—had had its first innings. Before the close of the year MM. Roberts and Charles constructed and inflated a hydrogen balloon, this time fitted with a practicable valve, and in partnership accomplished an ascent beating all previous records. The day, December 17, was one of winter temperature; yet the aeronauts quickly reached 6,000 feet, and when, after remaining aloft for one and a half hours, they descended, Roberts got out, leaving Charles in sole possession. Left to himself, this young recruit seems to have met with experiences which are certainly unusual, and which must be attributed largely to the novelty of his situation. He declared that at 9,000 feet, or less than two miles, all objects on the earth had disappeared from view, a statement which can only be taken to mean that he had entered cloud. Further, at this moderate elevation he not only became benumbed with cold, but felt severe pain in his right ear and jaw. He held on, however, ascending till 10,500 feet were reached, when he descended, having made a journey of thirty miles from the start.

Ascents, all on the Continent, now followed one another in rapid succession, and shortly the MM. Roberts essayed a venture on new lines. They attempted the guidance of a balloon by means of oars, and though they failed in this they were fortunate in making a fresh record. They also encountered a thunderstorm, and by adopting a perfectly scientific method—of which more hereafter—succeeded in eluding it. The storm broke around them when they were 14,000 feet high, and at this altitude, noting that there were diverse currents aloft, they managed to manoeuvre their balloon higher or lower at will and to suit their purpose, and by this stratagem drew away from the storm centre. After six and a half hours their voyage ended, but not until 150 miles had been covered.

It must be freely granted that prodigious progress had been made in an art that as yet was little more than a year old; but assuredly not enough to justify the absurdly inflated ideas that the Continental public now began to indulge in. Men lost their mental balance, allowing their imagination to run riot, and speculation became extravagant in the extreme. There was to be no limit henceforward to the attainment of fresh knowledge, nor any bounds placed to where man might roam. The universe was open to him: he might voyage if he willed to the moon or elsewhere: Paris was to be the starting point for other worlds: Heaven itself had been taken by storm.

Moderation had to be learned ere long by the discipline of more than one stern lesson. Hitherto a marvellous—call it a Providential—good fortune had attended the first aerial travellers; and even when mishaps presently came to be reckoned with, it may fairly be questioned whether so many lives were sacrificed among those who sought to voyage through the sky as were lost among such as first attempted to navigate the sea.

It is in such ventures as we are now regarding that fortune seems readiest to favour the daring, and if I may digress briefly to adduce experiences coming within my own knowledge, I would say that it is to his very impulsiveness that the enthusiast often owes the safety of his neck. It is the timid, not the bold rider, that comes to grief at the fence. It is the man who draws back who is knocked over by a tramcar. Sheer impetus, moral or physical, often carries you through, as in the case of a fall from horse-back. To tumble off when your horse is standing still and receive a dead blow from the ground might easily break a limb. But at full gallop immunity often lies in the fact that you strike the earth at an angle, and being carried forward, impact is less abrupt. I can only say that I have on more than one occasion found the greatest safety in a balloon venture involving the element of risk to lie in complete abandonment to circumstances, and in the increased life and activity which the delirium of excitement calls forth. In comparing, however, man's first ventures by sky with those by sea, we must remember what far greater demand the former must have made upon the spirit of enterprise and daring.

We can picture the earliest sea voyager taking his first lesson astride of a log with one foot on the bottom, and thus proceeding by sure stages till he had built his coracle and learned to paddle it in shoal water. But the case was wholly different when the first frail air ship stood at her moorings with straining gear and fiercely burning furnace, and when the sky sailor knew that no course was left him but to dive boldly up into an element whence there was no stepping back, and separated from earth by a gulf which man instinctively dreads to look down upon.

Taking events in their due sequence, we have now to record a voyage which the terrors of sky and sea together combined to make memorable. Winter had come—early January of 1785—when, in spite of short dark days and frosty air, M. Blanchard, accompanied by an American, Dr. Jeffries, determined on an attempt to cross the Channel. They chose the English side, and inflating their balloon with hydrogen at Dover, boldly cast off, and immediately drifted out to sea. Probably they had not paid due thought to the effect of low sun and chilly atmosphere, for their balloon rose sluggishly and began settling down ere little more than a quarter of their course was run. Thereupon they parted with a large portion of their ballast, with the result that they crept on as far as mid-Channel, when they began descending again, and cast out the residue of their sand, together with some books, and this, too, with the uncomfortable feeling that even these measures would not suffice to secure their safety.

This was in reality the first time that a sea passage had been made by sky, and the gravity of their situation must not be under-estimated. We are so accustomed in a sea passage to the constant passing of other vessels that we allow ourselves to imagine that a frequented portion of the ocean, such as the Channel, is thickly dotted over with shipping of some sort. But in entertaining this idea we are forgetful of the fact that we are all the while on a steamer track. The truth, however, is that anywhere outside such a track, even from the commanding point of view of a high-flying balloon, the ocean is seen to be more vast than we suppose, and bears exceedingly little but the restless waves upon its surface. Once fairly in the water with a fallen balloon, there is clearly no rising again, and the life of the balloon in this its wrong element is not likely to be a long one. The globe of gas may under favourable circumstances continue to float for some while, but the open wicker car is the worst possible boat for the luckless voyagers, while to leave it and cling to the rigging is but a forlorn hope, owing to the mass of netting which surrounds the silk, and which would prove a death-trap in the water. There are many instances of lives having been lost in such a dilemma, even when help was near at hand.

Our voyagers, whom we left in mid-air and stream, were soon descending again, and this time they threw out their tackle—anchor, ropes, and other gear, still without adequately mending matters. Then their case grew desperate. The French coast was, indeed, well in sight, but there seemed but slender chance of reaching it, when they began divesting themselves of clothing as a last resort. The upshot of this was remarkable, and deserves a moment's consideration. When a balloon has been lightened almost to the utmost the discharge of a small weight sometimes has a magical effect, as is not difficult to understand. Throwing out ten pounds at an early stage, when there may be five hundred pounds more of superfluous weight, will tell but little, but when those five hundred pounds are expended then an extra ten pounds scraped together from somewhere and cast overboard may cause a balloon to make a giant stride into space by way of final effort; and it was so with M. Blanchard. His expiring balloon shot up and over the approaching land, and came safely to earth near the Forest of Guiennes. A magnificent feast was held at Calais to celebrate the above event. M. Blanchard was presented with the freedom of the city in a gold box, and application was made to the Ministry to have the balloon purchased and deposited as a memorial in the church. On the testimony of the grandson of Dr. Jeffries the car of this balloon is now in the museum of the same city.

A very noteworthy example of how a balloon may be made to take a fresh lease of life is supplied by a voyage of M. Testu about this date, which must find brief mention in these pages. In one aspect it is laughable, in another it is sublime. From every point of view it is romantic.

It was four o'clock on a threatening day in June when the solitary aeronaut took flight from Paris in a small hydrogen balloon only partially filled, but rigged with some contrivance of wings which were designed to render it self-propelling. Discovering, however, that this device was inoperative, M. Testu, after about an hour and a half, allowed the balloon to descend to earth in a corn field, when, without quitting hold of the car, he commenced collecting stones for ballast. But as yet he knew not the ways of churlish proprietors of land, and in consequence was presently surprised by a troublesome crowd, who proceeded, as they supposed, to take him prisoner till he should pay heavy compensation, dragging him off to the nearest village by the trail rope of his balloon.

M. Testu now had leisure to consider his situation, and presently hit on a stratagem the like of which has often since been adopted by aeronauts in like predicament. Representing to his captors that without his wings he would be powerless, he suffered them to remove these weighty appendages, when also dropping a heavy cloak, he suddenly cut the cord by which he was being dragged, and, regaining freedom, soared away into the sky. He was quickly high aloft, and heard thunder below him, soon after which, the chill of evening beginning to bring him earthward, he descried a hunt in full cry, and succeeded in coming down near the huntsmen, some of whom galloped up to him, and for their benefit he ascended again, passing this time into dense cloud with thunder and lightning. He saw the sun go down and the lightning gather round, yet with admirable courage he lived the night out aloft till the storms were spent and the midsummer sun rose once more. With daylight restored, his journey ended at a spot over sixty miles from Paris.

We have, of course, recounted only a few of the more noteworthy early ballooning ventures. In reality there had up to the present time been scores of ascents made in different localities and in all conditions of wind and weather, yet not a life had been lost. We have now, however, to record a casualty which cost the first and boldest aeronaut his life, and which is all the more regrettable as being due to circumstances that should never have occurred.

M. Pilatre de Rosier, accompanied by M. Romain, determined on crossing the Channel from the French side; and, thinking to add to their buoyancy and avoid the risk of falling in the sea, hit on the extraordinary idea of using a fire balloon beneath another filled with hydrogen gas! With this deadly compound machine they actually ascended from Boulogne, and had not left the land when the inevitable catastrophe took place.

The balloons caught fire and blew up at a height of 3,000 feet, while the unfortunate voyagers were dashed to atoms.

CHAPTER III. THE FIRST BALLOON ASCENT IN ENGLAND.

As may be supposed, it was not long before the balloon was introduced into England. Indeed, the first successful ascent on record made in our own country took place in the summer of 1784, ten months previous to the fatal venture narrated at the close of the last chapter. Now, it is a remarkable and equally regrettable circumstance that though the first ascent on British soil was undoubtedly made by one of our own countrymen, the fact is almost universally forgotten, or ignored, and the credit is accorded to a foreigner.

Let us in strict honesty examine into the case. Vincent Lunardi, an Italian, Secretary to the Neapolitan Ambassador, Prince Caramanico, being in England in the year 1784, determined on organising and personally executing an ascent from London; and his splendid enterprise, which was presently carried to a successful issue, will form the principal subject of the present chapter. It will be seen that remarkable success crowned his efforts, and that his first and ever memorable voyage was carried through on September 15th of that year.

More than a month previously, however, attention had been called to the fact that a Mr. Tytler was preparing to make an ascent from Edinburgh in a hot air balloon, and in the London Chronicle of August 27th occurs the following circumstantial and remarkable letter from a correspondent to that journal:

"Edinburgh, Aug. 27, 1784.

"Mr. Tytler has made several improvements upon his fire balloon. The reason of its failure formerly was its being made of porous linen, through which the air made its escape. To remedy this defect, Mr. Tytler has got it covered with a varnish to retain the inflammable air after the balloon is filled.

"Early this morning this bold adventurer took his first aerial flight. The balloon being filled at Comely Garden, he seated himself in the basket, and the ropes being cut he ascended very high and descended quite gradually on the road to Restalrig, about half a mile from the place where he rose, to the great satisfaction of those spectators who were present. Mr. Tytler went up without the furnace this morning; when that is added he will be able to feed the balloon with inflammable air, and continue his aerial excursions as long as he chooses.

"Mr. Tytler is now in high spirits, and in his turn laughs at those infidels who ridiculed his scheme as visionary and impracticable. Mr. Tytler is the first person in Great Britain who has navigated the air."

Referring to this exploit, Tytler, in a laudatory epistle addressed to Lunardi, tells of the difficulties he had had to contend with, and artlessly reveals the cool, confident courage he must have displayed. No shelter being available for the inflation, and a strong wind blowing, his first misfortune was the setting fire to his wicker gallery. The next was the capsizing and damaging of his balloon, which he had lined with paper. He now substituted a coat of varnish for the paper, and his gallery being destroyed, so that he could no longer attempt to take up a stove, he resolved to ascend without one. In the end the balloon was successfully inflated, when he had the hardihood to entrust himself to a small basket (used for carrying earthenware) slung below, and thus to launch himself into the sky. He did so under the conviction that the risk he ran was greater than it really was, for he argued that his craft was now only like a projectile, and "must undoubtedly come to the ground with the same velocity with which it ascended." On this occasion the crowd tried for some time to hold him near the ground by one of the restraining ropes, so that his flight was curtailed. In a second experiment, however, he succeeded in rising some hundreds of feet, and came to earth without mishap.

But little further information respecting Mr. Tytler is apparently forthcoming, and therefore beyond recording the fact that he was the first British aeronaut, and also that he was the first to achieve a balloon ascent in Great Britain, we are unable to make further mention of him in this history.

Of his illustrious contemporary already mentioned there is, on the contrary, much to record, and we would desire to give full credit to his admirable courage and perseverance. It was with a certain national and pardonable pride that the young Italian planned his bold exploit, feeling with a sense of self-satisfaction, which he is at no pains to hide, that he aimed at winning honour for his country as well as for himself. In a letter which he wrote to his guardian, Chevalier Gherardo Compagni, he alludes to the stolid indifference of the English people and philosophers to the brilliant achievements in aeronautics which had been made and so much belauded on the Continent. He proclaims the rivalry as regards science and art existing between France and England, attributing to the latter an attitude of sullen jealousy. At the same time he is fully alive to the necessity of gaining English patronage, and sets about securing this with tactful diplomacy. First he casts about for a suitable spot where his enterprise would not fail to enlist general attention and perhaps powerful patrons, and here he is struck by the attractions and facilities offered by Chelsea Hospital. He therefore applies to Sir George Howard, the Governor, asking for the use of the famous hospital, to which, on the occasion of his experiments, he desires that admittance should only be granted to subscribers, while any profits should be devoted to the pensioners of the hospital. His application having been granted, he assures his guardian that he "still maintains his mental balance, and his sleep is not banished by the magnitude of his enterprise, which is destined to lead him through the path of danger to glory."

This letter was dated the 15th of July, and by the beginning of August his advertisement was already before the public, inviting subscribers and announcing a private view of his balloon at the Lyceum, where it was in course of construction, and was being fitted with contrivances of his own in the shape of oars and sails. He had by this time not only enlisted the interest of Sir George Howard, and of Sir Joseph Banks, but had secured the direct patronage of the King.

But within a fortnight a most unforeseen mishap had occurred, which threatened to overwhelm Lunardi in disappointment and ruin. A Frenchman of the name of Moret, designing to turn to his own advertisement the attention attracted by Lunardi's approaching trials, attempted to forestall the event by an enterprise of his own, announcing that he would make an ascent with a hot air balloon in some gardens near Chelsea Hospital, and at a date previous to that fixed upon by Lunardi. In attempting, however, to carry out this unworthy project the adventurer met with the discomfiture he deserved. He failed to effect his inflation, and when after fruitless attempts continued for three hours, his balloon refused to rise, a large crowd, estimated at 60,000, assembled outside, broke into the enclosure, committing havoc on all sides, not unattended with acts of violence and robbery.

The whole neighbourhood became alarmed, and it followed as a matter of course that Lunardi was peremptorily ordered to discontinue his preparations, and to announce in the public press that his ascent from Chelsea Hospital was forbidden. Failure and ruin now stared the young enthusiast in the face, and it was simply the generous feeling of the British public, and the desire to see fair play, that gave him another chance. As it was, he became the hero of the hour; thousands flocked to the show rooms at the Lyceum, and he shortly obtained fresh grounds, together with needful protection for his project, at the hands of the Hon. Artillery Company. By the 15th of September all incidental difficulties, the mere enumeration of which would unduly swell these pages, had been overcome by sheer persistence, and Lunardi stood in the enclosure allotted him, his preparations in due order, with 150,000 souls, who had formed for hours a dense mass of spectators, watching intently and now confidently the issue of his bold endeavour.

But his anxieties were as yet far from over, for a London crowd had never yet witnessed a balloon ascent, while but a month ago they had seen and wreaked their wrath upon the failure of an adventurer. They were not likely to be more tolerant now. And when the advertised hour for departure had arrived, and the balloon remained inadequately inflated, matters began to take a more serious turn. Half an hour later they approached a crisis, when it began to be known that the balloon still lacked buoyancy, and that the supply of gas was manifestly insufficient. The impatience of the mob indeed was kept in restraint by one man alone. This man was the Prince of Wales who, refusing to join the company within the building and careless of the attitude of the crowd, remained near the balloon to check disorder and unfair treatment.

But an hour after time the balloon still rested inert and then, with fine resolution, Lunardi tried one last expedient. He bade his colleague, Mr. Biggen, who was to have ascended with him, remain behind, and quietly substituting a smaller and lighter wicker car, or rather gallery, took his place within and severed the cords just as the last gun fired. The Prince of Wales raised his hat, imitated at once by all the bystanders, and the first balloon that ever quitted English soil rose into the air amid the extravagant enthusiasm of the multitude. The intrepid aeronaut, pardonably excited, and fearful lest he should not be seen within the gallery, made frantic efforts to attract attention by waving his flag, and worked his oars so vigorously that one of them broke and fell. A pigeon also gained its freedom and escaped. The voyager, however, still retained companions in his venture—a dog and a cat.

Following his own account, Lunardi's first act on finding himself fairly above the town was to fortify himself with some glasses of wine, and to devour the leg of a chicken. He describes the city as a vast beehive, St. Paul's and other churches standing out prominently; the streets shrunk to lines, and all humanity apparently transfixed and watching him. A little later he is equally struck with the view of the open country, and his ecstasy is pardonable in a novice. The verdant pastures eclipsed the visions of his own lands. The precision of boundaries impressed him with a sense of law and order, and of good administration in the country where he was a sojourner.

By this time he found his balloon, which had been only two-thirds full at starting, to be so distended that he was obliged to untie the mouth to release the strain. He also found that the condensed moisture round the neck had frozen. These two statements point to his having reached a considerable altitude, which is intelligible enough. It is, however, difficult to believe his further assertion that by the use of his single oar he succeeded in working himself down to within a few hundred feet of the earth. The descent of the balloon must, in point of fact, have been due to a copious outrush of gas at his former altitude. Had his oar really been effective in working the balloon down it would not have needed the discharge of ballast presently spoken of to cause it to reascend. Anyhow, he found himself sufficiently near the earth to land a passenger who was anxious to get out. His cat had not been comfortable in the cold upper regions, and now at its urgent appeal was deposited in a corn field, which was the point of first contact with the earth. It was carefully received by a country-woman, who promptly sold it to a gentleman on the other side of the hedge, who had been pursuing the balloon.

The first ascent of a balloon in England was deserving of some record, and an account alike circumstantial and picturesque is forthcoming. The novel and astonishing sight was witnessed by a Hertfordshire farmer, whose testimony, published by Lunardi in the same year, runs as follows: —

This deponent on his oath sayeth that, being on Wednesday, the 15th day of September instant, between the hours of three and four in the afternoon, in a certain field called Etna, in the parish of North Mimms aforesaid, he perceived a large machine sailing in the air, near the place where he was on horseback; that the machine continuing to approach the earth, the part of it in which this deponent perceived a gentleman standing came to the ground and dragged a short way on the ground in a slanting direction; that the time when this machine thus touched the earth was, as near as this deponent could judge, about a quarter before four in the afternoon. That this deponent being on horseback, and his horse restive, he could not approach nearer to the machine than about four poles, but that he could plainly perceive therein gentleman dressed in light coloured cloaths, holding in his hand a trumpet, which had the appearance of silver or bright tin. That by this time several harvest men coming up from the other part of the field, to the number of twelve men and thirteen women, this deponent called to them to endeavour to stop the machine, which the men attempted, but the gentleman in the machine desiring them to desist, and the machine moving with considerable rapidity, and clearing the earth, went off in a north direction and continued in sight at a very great height for near an hour afterwards. And this deponent further saith that the part of the machine in the which the gentleman stood did not actually touch the ground for more than half a minute, during which time the gentleman threw out a parcel of what appeared to this deponent as dry sand. That after the machine had ascended again from the earth this deponent perceived a grapple with four hooks, which hung from the bottom of the machine, dragging along the ground, which carried up with it into the air a small parcel of loose oats, which the women were raking in the field. And this deponent further on his oath sayeth that when the machine had risen clear from the ground about twenty yards the gentleman spoke to this deponent and to the rest of the people with his trumpet, wishing them goodbye and saying that he should soon go out of sight. And this deponent further on his oath sayeth that the machine in which the gentleman came down to earth appeared to consist of two distinct parts connected together by ropes, namely that in which the gentleman appeared to be, a stage boarded at the bottom, and covered with netting and ropes on the sides about four feet and a half high, and the other part of the machine appeared in the shape of an urn, about thirty feet high and of about the same diameter, made of canvas like oil skin, with green, red, and yellow stripes.

NATHANIEL WHITBREAD.

Sworn before me this twentieth day of September, 1784, WILLIAM BAKER.

It was a curious fact, pointed out to the brave Italian by a resident, that the field in which the temporary descent had been made was called indifferently Etna or Italy, "from the circumstance which attended the late enclosure of a large quantity of roots, rubbish, etc., having been collected there, and having continued burning for many days. The common people having heard of a burning mountain in Italy gave the field that name."

But the voyage did not end at Etna. The, as yet, inexperienced aeronaut now cast out all available ballast in the shape of sand, as also his provisions, and rising with great speed, soon reached a greater altitude than before, which he sought to still farther increase by throwing down his plates, knives, and forks. In this somewhat reckless expenditure he thought himself justified by the reliance he placed on his oar, and it is not surprising that in the end he owns that he owed his safety in his final descent to his good fortune. The narrative condensed concludes thus:—

"At twenty minutes past four I descended in a meadow near Ware. Some labourers were at work in it. I requested their assistance, but they exclaimed they would have nothing to do with one who came on the Devil's Horse, and no entreaties could prevail on them to approach me. I at last owed my deliverance to a young woman in the field who took hold of a cord I had thrown out, and, calling to the men, they yielded that assistance at her request which they had refused to mine."

As may be supposed, Lunardi's return to London resembled a royal progress. Indeed, he was welcomed as a conqueror to whom the whole town sought to do honour, and perhaps his greatest gratification came by way of the accounts he gathered of incidents which occurred during his eventful voyage. At a dinner at which he was being entertained by the Lord Mayor and judges he learned that a lady seeing his falling oar, and fancying that he himself was dashed to pieces, received a shock thereby which caused her death. Commenting on this, one of the judges bade him be reassured, inasmuch as he had, as if by compensation, saved the life of a young man who might live to be reformed. The young man was a criminal whose condemnation was regarded as certain at the hands of the jury before whom he was being arraigned, when tidings reached the court that Lunardi's balloon was in the air. On this so much confusion arose that the jury were unable to give due deliberation to the case, and, fearing to miss the great sight, actually agreed to acquit the prisoner, that they themselves might be free to leave the court!

But he was flattered by a compliment of a yet higher order. He was told that while he hovered over London the King was in conference with his principal Ministers, and his Majesty, learning that he was in the sky, is reported to have said to his councillors, "We may resume our own deliberations at pleasure, but we may never see poor Lunardi again!" On this, it is further stated that the conference broke up, and the King, attended by

Mr. Pitt and other chief officers of State, continued to view Lunardi through telescopes as long as he remained in the horizon.

The public Press, notably the Morning Post of September 16, paid a worthy tribute to the hero of the hour, and one last act of an exceptional character was carried out in his honour, and remains in evidence to this hour. In a meadow in the parish of Standon, near Ware, there stands a rough hewn stone, now protected by an iron rail. It marks the spot where Lunardi landed, and on it is cut a legend which runs thus:

> Let Posterity know And knowing be astonished that On the 15th day of September 1784 Vincent Lunardi of Lusca in Tuscany The first aerial traveller in Britain Mounting from the Artillery Ground In London And Traversing the Regions of the Air For Two Hours and Fifteen Minutes In this Spot Revisited the Earth. On this rude monument For ages be recorded That Wondrous Enterprise Successfully atchieved By the Powers of Chemistry And the Fortitude of Man That Improvement in Science Which The Great Author of all Knowledge Patronyzing by His Providence The Invention of Mankind Hath

graciously permitted To Their Benefit And
His own Eternal Glory.

CHAPTER IV. THE DEVELOPMENT OF BALLOON PHILOSOPHY.

In less than two years not only had the science of ballooning reached almost its highest development, but the balloon itself, as an aerostatic machine, had been brought to a state of perfection which has been but little improved upon up to the present hour. Better or cheaper methods of inflation were yet to be discovered, lighter and more suitable material remained to be manufactured; but the navigation of the air, which hitherto through all time had been beyond man's grasp, had been attained, as it were, at a bound, and at the hands of many different and independent experimentalists was being pursued with almost the same degree of success and safety as to-day.

Nor was this all. There was yet another triumph of the aeronautical art which, within the same brief period, had been to all intents and purposes achieved, even if it had not been brought to the same state of perfection as at the present hour. This was the Parachute. This fact is one which for a sufficient reason is not generally known. It is very commonly supposed that the parachute, in anything like its present form, is a very modern device, and that the art of successfully using it had not been introduced to the world even so lately as thirty years ago. Thus, we find it stated in works of that date dealing with the subject that disastrous consequences almost necessarily attended the use of the parachute, "the defects of which had been attempted to be remedied in various ways, but up to this time without success." A more correct statement, however, would have been that the art of constructing and using a practicable parachute had through many years been lost or forgotten. In actual fact, it had been adopted with every assurance of complete success by the year 1785, when Blanchard by its means lowered dogs and other animals with safety from a balloon. A few years later he descended himself in a like apparatus from Basle, meeting, however, with the misadventure of a broken leg.

But we must go much further back for the actual conception of the parachute, which, we might suppose, may originally have been suggested by the easy floating motion with which certain seeds or leaves will descend from lofty trees, or by the mode adopted by birds of dropping softly to earth with out-stretched wings. M. de la Loubere, in his historical account of Siam, which he visited in 1687-88, speaks of an ingenious athlete who exceedingly diverted the King and his court by leaping from a height and supporting himself in the air by two umbrellas, the handles of which were affixed to his girdle. In 1783, that is, the same year as that in which the balloon was invented, M. le Normand experimented with a like umbrella-shaped contrivance, with a view to its adoption as a fire escape, and he demonstrated the soundness of the principle by descending himself from the windows of a lofty house at Lyons.

It was, however, reserved for M. Jacques Garnerin in 1797 to make the first parachute descent that attracted general attention. Garnerin had previously been detained as a State prisoner in the fortress of Bade, in Hungary, after the battle of Marchiennes in 1793, and during his confinement had pondered on the possibility of effecting his escape by a parachute. His solitary cogitations and calculations resulted, after his release, in the invention and construction of an apparatus which he put to a practical test at Paris before the court of France on October 22nd, 1797. Ascending in a hydrogen balloon to the height of about 2,000 feet, he unhesitatingly cut himself adrift, when for some distance he dropped like a stone. The folds of his apparatus, however, opening suddenly, his fall became instantly checked. The remainder of his descent, though leisurely, occupying, in fact, some twelve minutes, appeared to the spectators to be attended with uncertainty, owing to a swinging motion set up in the car to which he was clinging. But the fact remains that he reached the earth with only slight impact, and entirely without injury.

It appears that Garnerin subsequently made many equally successful parachute descents in France, and during the short peace of 1802 visited London, where he gave an exhibition of his art. From the most reliable accounts of his exploit it would seem that his drop was from a very great height, and that a strong ground wind was blowing at the time, the result of which was that wild, wide oscillations were set up in the car, which narrowly escaped bringing him in contact with the house tops in St. Pancreas, and eventually swung him down into a field, not without some unpleasant scratches.

Nor was Garnerin the only successful parachutist at this period. A Polish aeronaut, Jordaki Kuparento, ascended from Warsaw on the 24th of July, 1804 in a hot air balloon, taking up, as was the custom, an attached furnace, which caused the balloon to take fire when at a great height. Kuparento, however, who was alone, had as a precaution provided himself with a parachute, and with this he seems to have found no difficulty in effecting a safe descent to earth.

It was many years after this that fresh experimentalists, introducing parachutes on new lines and faulty in construction, met with death or disaster. Enough, however, has already been said to show that in the early years we are now traversing in this history a perfectly practicable parachute had become an accomplished fact. The early form is well described by Mr. Monck Mason in a letter to the Morning Herald in 1837, written on the eve of an unrehearsed and fatal experiment made by Mr. Cocking, which must receive notice in due course. "The principle," writes Mr. Monck Mason, "upon which all these parachutes were constructed is the same, and consists simply of a flattened dome of silk or linen from 24 feet to 28 feet in diameter. From the outer margin all around at stated intervals proceed a large number of cords, in length about the diameter of the dome itself, which, being collected together in one point and made fast to another of superior dimensions attached to the apex of the machine, serve to maintain it in its form when expanded in the progress of the descent. To this centre cord likewise, at a distance below the point of junction, varying according to the fancy of the aeronaut, is fixed the car or basket in which he is seated, and the whole suspended from the network of the balloon in such a manner as to be capable of being detached in an instant at the will of the individual by cutting the rope by which it is made fast above."

It followed almost as a matter of course that so soon as the balloon had been made subject to something like due control, and thus had become recognised as a new machine fairly reduced to the service of man, it began to be regarded as an instrument which should be made capable of being devoted to scientific research. Indeed, it may be claimed that, among the very earliest aeronauts, those who had sailed away into the skies and brought back intelligent observations or impressions of the realm of cloud-land, or who had only described their own sensations at lofty altitudes, had already contributed facts of value to science. It is time then, taking events in their due sequence, that mention should be made of the endeavours of various savants, who began about the commencement of the nineteenth century to gather fresh knowledge from the exploration of the air by balloon ascents organised with fitting equipment. The time had now come for promoting the balloon to higher purposes than those of mere exhibition or amusement. In point of fact, it had already in one way been turned to serious practical account. It had been used by the French during military operations in the revolutionary war as a mode of reconnoitring, and not without success, so that when after due trial the war balloon was judged of value a number of similar balloons were constructed for the use of the various divisions of the French army, and, as will be told in its proper place, one, at least, of these was put to a positive test before the battle of Fleurus.

But, returning to more strictly scientific ascents, which began to be mooted at this period, we are at once impressed with the widespread influence which the balloon was exercising on thinking minds. We note this from the fact that what must be claimed to be the first genuine ascent for scientific observation was made in altogether fresh ground, and at so distant a spot as St. Petersburg.

It was now the year 1804, and the Russian Academy had determined on attempting an examination of the physical condition of the higher atmosphere by means of the balloon. The idea had probably been suggested by scientific observations which had already been made on mountain heights by such explorers as De Luc, Saussure, Humboldt, and others. And now it was determined that their results should be tested alongside such observations as could be gathered in the free heaven far removed from any disturbing effects that might be caused by contiguity to earth. The lines of enquiry to which special atten-

tion was required were such as would be naturally suggested by the scientific knowledge of the
hour, though they may read somewhat quaintly
to-day. Would there be any change in the intensity of the magnetic force? Any change in the inclination of the magnetised needle? Would evaporation find a new law? Would solar rays increase in
power? What amount of electric matter would be
found? What change in the colours produced by
the prism? What would be the constitution of the
higher and more attenuated air? What physical
effect would it have on human and bird life?

The ascent was made at 7.15 on a summer evening by M. Robertson and the Academician, M.
Sacharof, to whom we are indebted for the following resume of notes, which have a special
value as being the first of their class. Rising slowly, a difference of atmosphere over the Neva gave
the balloon a downward motion, necessitating
the discharge of ballast. As late as 8.45 p.m. a fine
view was obtained of the Newski Islands, and the
whole course of the neighbouring river. At 9.20
p.m., when the barometer had fallen from 30
inches to 23 inches, a canary and a dove were
dismissed, the former falling precipitately, while
the latter sailed down to a village below. All
available ballast was now thrown out, including a
spare great coat and the remains of supper, with

the result that at 9.30 the barometer had fallen to 22 inches, and at this height they caught sight of the upper rim of the sun. The action of heart and lungs remained normal. No stars were seen, though the sky was mainly clear, such clouds as were visible appearing white and at a great height. The echo of a speaking trumpet was heard after an interval of ten seconds. This was substantially the outcome of the experiments. The practical difficulties of carrying out prearranged observations amid the inconvenience of balloon travel were much felt. Their instruments were seriously damaged, and their results, despite most painstaking and praiseworthy efforts, must be regarded as somewhat disappointing.

But ere the autumn of the same year two other scientific ascents, admirably schemed and financed at the public expense, had been successfully carried out at Paris in a war balloon which, as will be told, had at this time been returned from military operations in Egypt. In the first of these, Gay Lussac ascended in company with M. Biot, with very complete equipment. Choosing ten o'clock in the morning for their hour of departure, they quickly entered a region of thin, but wet fog, after which they shot up into denser cloud, which they completely surmounted at a height of 6,500 feet, when they described the upper surface as bearing the resemblance, familiar enough to aeronauts and mountaineers, as of a white sea broken up into gently swelling billows, or of an extended plain covered with snow.

A series of simple experiments now embarked upon showed the behaviour of magnetised iron, as also of a galvanic pile or battery, to remain unaltered. As their altitude increased their pulses quickened, though beyond feeling keenly the contrast of a colder air and of scorching rays of the sun they experienced no physical discomfort. At 11,000 feet a linnet which they liberated fell to the earth almost helplessly, while a pigeon with difficulty maintained an irregular and precipitate flight. A carefully compiled record was made of variations of temperature and humidity, and they succeeded in determining that the upper air was charged with negative electricity. In all this these two accomplished physicists may be said to have carried out a brilliant achievement, even though their actual results may seem somewhat meagre. They not only were their own aeronauts, but succeeded in arranging and carrying out continuous and systematic observations throughout the period of their remaining in the sky.

This voyage was regarded as such a pronounced success that three weeks later, in mid-September, Gay Lussac was induced to ascend again, this time alone, and under circumstances that should enable him to reach an exceptionally high altitude. Experience had taught the advisability of certain modifications in his equipment. A magnet was ingeniously slung with a view of testing its oscillation even in spite of accidental gyrations in the balloon. Thermometers and hygrometers were carefully sheltered from the direct action of the sun, and exhausted flasks were supplied with the object of bringing down samples of upper air for subsequent analysis.

Again it was an early morning ascent, with a barometer on the ground standing at 30.6 inches, and a slightly misty air. Lussac appears to have accomplished the exceedingly difficult task of counting the oscillations of his magnet with satisfaction to himself. At 10,000 feet twenty vibrations occupied 83 seconds, as compared with 84.33 seconds at the earth's surface. The variation of the compass remained unaltered, as also the behaviour of magnetised iron at all altitudes. Keeping his balloon under perfect control, and maintaining a uniform and steady ascent, he at the same time succeeded in compiling an accurate table of readings recording atmospheric pressure, temperature and humidity, and it is interesting to find that he was confronted with an apparent anomaly which will commonly present itself to the aeronaut observer. Up to 12,000 feet the temperature had decreased consistently from 82 degrees to 47 degrees, after which it increased 6 degrees in the next 2,000 feet. This by no means uncommon experience shall be presently discussed. The balloon was now steadily manoeuvred up to 18,636 feet, at which height freezing point was practically reached. Then with a further climb 20,000 feet is recorded, at which altitude the ardent philosopher could still attend to his magnetic observations, nor is his arduous and unassisted task abandoned here, but with marvellous pertinacity he yet struggled upwards till a height of no less than 23,000 feet is recorded, and the thermometer had sunk to 14 degrees F. Four miles and a quarter above the level of the sea, reached by a solitary aerial explorer, whose legitimate training lay apart from aeronautics, and whose main care was the observation of the philosophical instruments he carried! The achievement of this French savant makes a brilliant record in the early pages of our history.

It is not surprising that Lussac should own to having felt no inconsiderable personal discomfort before his venture was over. In spite of warm clothing he suffered greatly from cold and benumbed fingers, not less also from laboured breathing and a quickened pulse; headache supervened, and his throat became parched and unable to swallow food. In spite of all, he conducted the descent with the utmost skill, climbing down quietly and gradually till he alighted with gentle ease at St. Gourgen, near Rouen. It may be mentioned here that the analysis of the samples of air which he had brought down proved them to contain the normal proportion of oxygen, and to be essentially identical, as tested in the laboratory, with the free air secured at the surface of the earth.

The sudden and apparently unaccountable variation in temperature recorded by Lussac is a striking revelation to an aerial observer, and becomes yet more marked when more sensitive instruments are used than those which were taken up on the occasion just related. It will be recorded in a future chapter how more suitable instruments came in course of time to be devised. It is only necessary to point out at this stage that instruments which lack due sensibility will unavoidably read too high in ascents, and too low in descents where, according to the general law, the air is found to grow constantly colder with elevation above the earth's surface. It is strong evidence of considerable efficiency in the instruments, and of careful attention on the part of the observer, that Lussac was able to record the temporary inversion of the law of change of temperature above-mentioned. Had he possessed modern instrumental equipment he would have brought down a yet more remarkable account of the upper regions which he visited, and learned that the variations of heat and cold were considerably more striking than he supposed.

With a specially devised instrument used with special precautions, the writer, as will be shown hereafter, has been able to prove that the temperature of the air, as traversed in the wayward course of a balloon, is probably far more variable and complex than has been recorded by most observers.

The exceptional height claimed to have been reached by Gay Lassac need not for a moment be questioned, and the fact that he did not experience the same personal inconvenience as has been complained of by mountain climbers at far less altitudes admits of ready explanation. The physical exertion demanded of the mountaineer is entirely absent in the case of an aeronaut who is sailing at perfect ease in a free balloon. Moreover, it must be remembered that—a most important consideration—the aerial voyager, necessarily travelling with the wind, is unconscious, save at exceptional moments, of any breeze whatever, and it is a well-established fact that a degree of cold which might be insupportable when a breeze is stirring may be but little felt in dead calm. It should also be remembered, in duly regarding Gay Lussac's remarkable record, that this was not his first experience of high altitudes, and it is an acknowledged truth that an aeronaut, especially if he be an enthusiast, quickly becomes acclimatised to his new element, and sufficiently inured to its occasional rigours.

CHAPTER V. SOME FAMOUS EARLY VOYAGERS.

During certain years which now follow it will possibly be thought that our history, so far as incidents of special interest are concerned, somewhat languishes. Yet it may be wrong to regard this period as one of stagnation or retrogression.

Before passing on to later annals, however, we must duly chronicle certain exceptional achievements and endeavours as yet unmentioned, which stand out prominently in the period we have been regarding as also in the advancing years of the new century Among these must in justice be included those which come into the remarkable, if somewhat pathetic subsequent career of the brilliant, intrepid Lunardi.

Compelling everywhere unbounded admiration he readily secured the means necessary for carrying out further exploits wherever he desired while at the same time he met with a measure of good fortune in freedom from misadventure such as has generally been denied to less bold adventurers. Within a few months of the time when we left him, the popular hero and happy recipient of civic and royal favours, we find him in Scotland attempting feats which a knowledge of practical difficulties bids us regard as extraordinary.

To begin with, nothing appears more remarkable than the ease, expedition, and certainty with which in days when necessary facilities must have been far harder to come by than now, he could always fill his balloon by the usually tedious and troublesome mode attending hydrogen inflation. We see him at his first Scottish ascent, completing the operation in little more than two hours. It is the same later at Glasgow, where, commencing with only a portion of his apparatus, he finds the inflation actually to proceed too rapidly for his purpose, and has to hold the powers at his command strongly in check. Later, in December weather, having still further improved his apparatus, he makes his balloon support itself after the inflation of only ten minutes. Then, as if assured of impunity, he treats recognised risks with a species of contempt. At Kelso he hails almost with joy the fact that the wind must carry him rapidly towards the sea, which in the end he narrowly escapes. At Glasgow the chances of safe landing are still more against him, yet he has no hesitation in starting, and at last the catastrophe he seemed to court actually overtook him, and he plumped into the sea near Berwick, where no sail was even in sight, and a winter's night coming on. From this predicament he was rescued by a special providence which once before had not deserted him, when in a tumult of violent and contrary currents, and at a great height to boot, his gallery was almost completely carried away, and he had to cling on to the hoop desperately with both hands.

Then we lose sight of the dauntless, light-hearted Italian for one-and-twenty years, when in the Gentleman's Magazine of July 31, 1806, appears the brief line, "Died in the convent of Barbadinas, of a decline, Mr. Vincent Lunardi, the celebrated aeronaut."

Garnerin, of whom mention has already been made, accomplished in the summer of 1802 two aerial voyages marked by extreme velocity in the rate of travel. The first of these is also remarkable as having been the first to fairly cross the heart of London. Captain Snowdon, R.N., accompanied the aeronaut. The ascent took place from Chelsea Gardens, and proved so great an attraction that the crowd overflowed into the neighbouring parts of the town, choking up the thoroughfares with vehicles, and covering the river with boats. On being liberated, the balloon sped rapidly away, taking a course midway between the river and the main highway of the Strand, Fleet Street, and Cheapside, and so passed from view of the multitude. Such a departure could hardly fail to lead to subsequent adventures, and this is pithily told in a letter written by Garnerin himself: "I take the earliest opportunity of informing you that after a very pleasant journey, but after the most dangerous descent I ever made, on account of the boisterous weather and the vicinity of the sea, we alighted at the distance of four miles from this place and sixty from Ranelagh. We were only three-quarters of an hour on the way. To-night I intend to be in London with the balloon, which is torn to pieces. We ourselves are all over bruises."

Only a week after the same aeronaut ascended again from Marylebone, when he attained almost the same velocity, reaching Chingford, a distance of seventeen miles, in fifteen minutes.

The chief danger attending a balloon journey in a high wind, supposing no injury has been sustained in filling and launching, results not so much from impact with the ground on alighting as from the subsequent almost inevitable dragging along the ground. The grapnels, spurning the open, will often obtain no grip save in a hedge or tree, and even then large boughs will be broken through or dragged away, releasing the balloon on a fresh career which may, for a while, increase in mad impetuosity as the emptying silk offers a deeper hollow for the wind to catch.

The element of risk is of another nature in the case of a night ascent, when the actual alighting ground cannot be duly chosen or foreseen. Among many record night ascents may here, somewhat by anticipation of events, be mentioned two embarked upon by the hero of our last adventure. M. Garnerin was engaged to make a spectacular ascent from Tivoli at Paris, leaving the grounds at night with attached lamps illuminating his balloon. His first essay was on a night of early August, when he ascended at 11 p.m., reaching a height of nearly three miles. Remaining aloft through the hours of darkness, he witnessed the sun rise at half-past two in the morning, and eventually came to earth after a journey of some seven hours, during which time he had covered considerably more than a hundred miles. A like bold adventure carried out from the same grounds the following month was attended with graver peril. A heavy thunderstorm appearing imminent, Garnerin elected to ascend with great rapidity, with the result that his balloon, under the diminished pressure, quickly became distended to an alarming degree, and he was reduced to the necessity of piercing a hole in the silk, while for safety's sake he endeavoured to extinguish all lamps within reach. He now lost all control over his balloon, which became unmanageable in the conflict of the storm. Having exhausted his ballast, he presently was rudely brought to earth and then borne against a mountain side, finally losing consciousness until the balloon had found anchorage three hundred miles away from Paris.

A night ascent, which reads as yet more sensational and extraordinary, is reported to have been made a year or two previously, and when it is considered that the balloon used was of the Montgolfier type the account as it is handed down will be allowed to be without parallel. It runs thus: Count Zambeccari, Dr. Grassati of Rome, and M. Pascal Andreoli of Antona ascended on a November night from Bologna, allowing their balloon to rise with excessive velocity. In consequence of this rapid transition to an extreme altitude the Count and the Doctor became insensible, leaving Andreoli alone in possession of his faculties. At two o'clock in the morning they found themselves descending over the Adriatic, at which time a lantern which they carried expired and was with difficulty re-lighted. Continuing to descend, they presently pitched in to the sea and became drenched with salt water. It may seem surprising that the balloon, which could not be prevented falling in the water, is yet enabled to ascend from the grip of the waves by the mere discharge of ballast. (It would be interesting to inquire what meanwhile happened to the fire which they presumably carried with them.) They now rose into regions of cloud, where they became covered with hoar frost and also stone deaf. At 3 a.m. they were off the coast of Istria, once more battling with the waves till picked up by a shore boat. The balloon, relieved of their weight, then flew away into Turkey.

However overdrawn this narrative may appear, it must be read in the light of another account, the bare, hard facts of which can admit of no question. It is five years later, and once again Count Zambeccari is ascending from Bologna, this time in company with Signor Bonagna. Again it is a Montgolfier or fire balloon, and on nearing earth it becomes entangled in a tree and catches fire. The aeronauts jump for their lives, and the Count is killed on the spot. Certainly, when every allowance is made for pardonable or unintentional exaggeration, it must be conceded that there were giants in those days. Giants in the conception and accomplishment of deeds of lofty daring. Men who came scathless through supreme danger by virtue of the calmness and courage with which they withstood it.

Among other appalling disasters we have an example of a terrific descent from a vast height in which the adventurers yet escape with their lives. It was the summer of 1808, and the aeronauts, MM. Andreoli and Brioschi, ascending from Padua, reach a height at which a barometer sinks to eight inches, indicating upwards of 30,000 feet. At this point the balloon bursts, and falls precipitately near Petrarch's tomb. Commenting on this, Mr. Glaisher, the value of whose opinion is second to none, is not disposed to question the general truth of the narrative. In regard to Zambeccari's escape from the sea related above, it should be stated that in the case of a gas-inflated balloon which has no more than dipped its car or gallery in the waves, it is generally perfectly possible to raise it again from the water, provided there is on board a store of ballast, the discharge of which will sufficiently lighten the balloon. A case in point occurred in a most romantic and perilous voyage accomplished by Mr. Sadler on the 1st of October, 1812.

His adventure is one of extraordinary interest, and of no little value to the practical aeronaut. The following account is condensed from Mr. Sadler's own narrative. He started from the grounds of Belvedere House, Dublin, with the expressed intention of endeavouring to cross over the Irish Channel to Liverpool. There appear to have been two principal air drifts, an upper and a lower, by means of which he entertained fair hopes of steering his desired course. But from the outset he was menaced with dangers and difficulties. Ere he had left the land he discovered a rent in his silk which, occasioned by some accident before leaving, showed signs of extending. To reach this, it was necessary to extemporise by means of a rope a species of ratlins by which he could climb the rigging. He then contrived to close the rent with his neckcloth. He was, by this time, over the sea, and, manoeuvring his craft by aid of the two currents at his disposal, he was carried to the south shore of the Isle of Man, whence he was confident of being able, had he desired it, of landing in Cumberland. This, however, being contrary to his intention, he entrusted himself to the higher current, and by it was carried to the north-west of Holyhead. Here he dropped once again to the lower current, drifting south of the Skerry Lighthouse across the Isle of Anglesea, and at 4.30 p.m. found himself abreast of the Great Orme's Head. Evening now approaching, he had determined to seek a landing, but at this critical juncture the wind shifted to the southward, and he became blown out to sea. Then, for an hour, he appears to have tried high and low for a more favourable current, but without success; and, feeling the danger of his situation, and, moreover, sighting no less than five vessels beating down the Channel, he boldly descended in the sea about a mile astern of them. He must for certain have been observed by these vessels; but each and all held on their course, and, thus deserted, the aeronaut had no choice but to discharge ballast, and, quitting the waves, to regain his legitimate element. His experiences at this period of his extraordinary voyage are best told in his own words. "At the time I descended the sun was near setting Already the shadows of evening had cast a dusky hue over the face of the ocean, and a crimson glow purpled the tops of the waves as, heaving in the evening breeze, they died away in distance, or broke in foam against the sides of the vessels, and before I rose from the sea the orb had sunk below the horizon, leaving only the twilight glimmer to light the vast expanse around me. How great, therefore, was my astonishment, and how incapable is expression to convey an adequate idea of my feelings when, rising to the upper region of the air, the sun, whose parting beams I had already witnessed, again burst on my view, and encompassed me with the full blaze of day. Beneath me hung the shadows of even, whilst the clear beams of the sun glittered on the floating vehicle which bore me along rapidly before the wind."

After a while he sights three more vessels, which signify their willingness to stand by, whereupon he promptly descends, dropping beneath the two rear-most of them. From this point the narrative of the sinking man, and the gallant attempt at rescue, will rival any like tale of the sea. For the wind, now fast rising, caught the half empty balloon so soon as the car touched the sea, and the vessel astern, though in full pursuit, was wholly unable to come up. Observing this, Mr. Sadler, trusting more to the vessel ahead, dropped his grappling iron by way of drag, and shortly afterwards tried the further expedient of taking off his clothes and attaching them to the iron. The vessels, despite these endeavours, failing to overhaul him, he at last, though with reasonable reluctance, determined to further cripple the craft that bore him so rapidly by liberating a large quantity of gas, a desperate, though necessary, expedient which nearly cost him his life.

For the car now instantly sank, and the unfortunate man, clutching at the hoop, found he could not even so keep himself above the water, and was reduced to clinging, as a last hope, to the netting. The result of this could be foreseen, for he was frequently plunged under water by the mere rolling of the balloon. Cold and exertion soon told on him, as he clung frantically to the valve rope, and when his strength failed him he actually risked the expedient of passing his head through the meshes of the net. It was obvious that for avail help must soon come; yet the pursuing vessel, now close, appeared to hold off, fearing to become entangled in the net, and in this desperate extremity, fainting from exhaustion and scarcely able to cry aloud, Mr. Sadler himself seems to have divined the chance yet left; for, summoning his failing strength, he shouted to the sailors to run their bowsprit through his balloon. This was done, and the drowning man was hauled on board with the life scarcely in him.

A fitting sequel to the above adventure followed five years afterwards. The Irish Sea remained unconquered. No balloonist had as yet ever crossed its waters. Who would attempt the feat once more? Who more worthy than the hero's own son, Mr. Windham Sadler?

This aspiring aeronaut, emulating his father's enterprising spirit, chose the same starting ground at Dublin, and on the longest day of 1817, when winds seemed favourable, left the Porto Bello barracks at 1.20 p.m. His endeavour was to "tack" his course by such currents as he should find, in the manner attempted by his father, and at starting the ground current blew favourably from the W.S.W. He, however, allowed his balloon to rise to too high an altitude, where he must have been taken aback by a contrary drift; for, on descending again through a shower of snow, he found himself no further than Ben Howth, as yet only ten miles on his long journey. Profiting by his mistake, he thenceforward, by skilful regulation, kept his balloon within due limits, and successfully maintained a direct course across the sea, reaching a spot in Wales not far from Holyhead an hour and a half before sundown. The course taken was absolutely the shortest possible, being little more than seventy miles, which he traversed in five hours.

From this period of our story, noteworthy events in aeronautical history grow few and far between. As a mere exhibition the novelty of a balloon ascent had much worn off. No experimentalist was ready with any new departure in the art. No fresh adventure presented itself to the minds of the more enterprising spirits; and, whereas a few years previously ballooning exploits crowded into every summer season and were not neglected even in winter months, there is now for a while little to chronicle, either abroad or in our own country. A certain revival of the sensational element in ballooning was occasionally witnessed, and not without mishap, as in the case of Madame Blanchard, who, in the summer of 1819, ascending at night with fireworks from the Tivoli Gardens, Paris, managed to set fire to her balloon and lost her life in her terrific fall. Half a dozen years later a Mr., as also Mrs., Graham figure before the public in some bold spectacular ascents.

But the fame of any aeronaut of that date must inevitably pale before the dawning light shed by two stars of the first magnitude that were arising in two opposite parts of the world—Mr. John Wise in America, and Mr. Charles Green in our own country. The latter of these, who has been well styled the "Father of English Aeronautics," now entered on a long and honoured career of so great importance and success that we must reserve for him a separate and special chapter.

CHAPTER VI. CHARLES GREEN AND THE NASSAU BALLOON.

The balloon, which had gradually been dropping out of favour, had now been virtually laid aside, and, to all appearance, might have continued so, when, as if by chance concurrence of events, there arrived both the hour and the man to restore it to the world, and to invest it with a new practicability and importance. The coronation of George the Fourth was at hand, and this became a befitting occasion for the rare genius mentioned at the end of the last chapter, and now in his thirty-sixth year, to put in practice a new method of balloon management and inflation, the entire credit of which must be accorded to him alone.

From its very introduction and inception the gas balloon, an expensive and fragile structure in itself, had proved at all times exceedingly costly in actual use. Indeed, we find that at the date at which we have now arrived the estimate for filling a balloon of 70,000 cubic feet—no extraordinary capacity—with hydrogen gas was about L250. When, then, to this great outlay was added the difficulty and delay of producing a sufficient supply by what was at best a clumsy process, as also the positive failure and consequent disappointment which not infrequently ensued, it is easy to understand how through many years balloon ascents, no longer a novelty, had begun to be regarded with distrust, and the profession of a balloonist was doomed to become unremunerative. A simpler and cheaper mode of inflation was not only a desideratum, but an absolute necessity. The full truth of this may be gathered from the fact that we find there were not seldom instances where two or three days of continuous and anxious labour were expended in generating and passing hydrogen into a balloon, through the fabric of which the subtle gas would escape almost as fast as it was produced.

It was at this juncture, then, that Charles Green conceived the happy idea of substituting for hydrogen gas the ordinary household gas, which at this time was to be found ready to hand and in sufficient quantity in all towns of any consequence; and by the day of the coronation all was in readiness for a public exhibition of this method of inflation, which was carried out with complete success, though not altogether without unrehearsed and amusing incident, as must be told.

The day, July 18, was one of summer heat, and Green at the conclusion of his preparations, fatigued with anxious labour and oppressed by the crowding of the populace, took refuge within the car of his balloon, which was by that time already inflated, and only awaiting the gun signal that was to announce the moment for its departure. To allow of his gaining the refreshment of somewhat purer air he begged his friends who were holding the car of his balloon in restraint to keep it suspended at a few feet from the earth, while he rested himself within, and, this being done, it would appear that he fell into a doze, from which he did not awake till he found that the balloon, which had slipped from his friends' hold, was already high above the crowd and requiring his prompt attention. This was, however, by no means an untoward accident, and Green's triumph was complete. By this one venture alone the success of the new method was entirely assured. The cost of the inflation had been reduced ten-fold, the labour and uncertainty a hundred-fold, and, over and above all, the confidence of the public was restored. It is little wonder, then, that in the years that now follow we find the balloon returning to all the favour it had enjoyed in its palmiest days. But Green proved himself something more than a practical balloonist of the first rank. He brought to the aid of his profession ideas which were matured by due thought and scientifically sound. It is true he still clung for a while to the antiquated notion that mechanical means could, with advantage, be used to cause a balloon to ascend or descend, or to alter its direction in a tranquil atmosphere. But he saw clearly that the true method of navigating a balloon should be by a study of upper currents, and this he was able to put to practical proof on a memorable occasion, and in a striking manner, as we shall presently relate.

He learned the lesson early in his career while acquiring facts and experience, unassisted, in a number of solitary voyages made from different parts of the country. Among these he is careful to record an occasion when, making a day-light ascent from Boston, Lincolnshire, he maintained a lofty course, which promised to take him direct to Grantham; but, presently descending to a lower level, and his balloon diverging at an angle of some 45 degrees, he now headed for Newark. This experience he stored away.

A month later we find him making a night voyage from Vauxhall Gardens, destined to be the scene of many memorable ascents in the near future; and on this occasion he gave proof of his capability as a close and intelligent observer. It was a July night, near 11 p.m., moonless and cloudy, yet the earth was visible, and under these circumstances his simple narrative becomes of scientific value. He accurately distinguished the reflective properties of the face of the diversified country he traversed. Over Battersea and Wandsworth—this was in 1826—there were white sheets spread over the land, which proved to be corn crops ready for the sickle. Where crops were not the ground was darker, with, here and there, objects absolutely black—in other words, trees and houses. Then he mentions the river in a memorandum, which reads strangely to the aeronaut who has made the same night voyage in these latter days. The stream was crossed in places with rows of lamps apparently resting on the water. These were the lighted bridges; but, here and there, were dark planks, and these too were bridges—at Battersea and Putney—but without a light upon them!

In these and many other simple, but graphic, narratives Green draws his own pictures of Nature in her quieter moods. But he was not without early experience of her horse play, a highly instructive record of which should not be omitted here, and which, as coming from so careful and conscientious an observer, is best gathered from his own words. The ascent was from Newbury, and it can have been no mean feat to fill, under ordinary circumstances, a balloon carrying two passengers and a considerable weight of ballast at the small gas-holder which served the town eighty-five years ago. But the circumstances were not ordinary, for the wind was extremely squally; a tremendous hail and thunderstorm blew up, and a hurricane swept the balloon with such force that two tons weight of iron and a hundred men scarce sufficed to hold it in check.

Green on this occasion had indeed a companion, whose usefulness however at a pinch may be doubted when we learn that he was both deaf and dumb. The rest of the narrative runs thus: "Between 4 and 5 p.m. the clouds dispersed, but the wind continued to rage with unabated fury the whole of the evening. At 6 p.m. I stepped into the car with Mr. Simmons and gave the word 'Away!' The moment the machine was disencumbered of its weights it was torn by the violence of the wind from the assistants, bounded off with the velocity of lightning in a southeasterly direction, and in a very short space of time attained an elevation of two miles. At this altitude we perceived two immense bodies of clouds operated on by contrary currents of air until at length they became united, and at that moment my ears were assailed by the most awful and longest continued peal of thunder I have ever heard. These clouds were a full mile beneath us, but perceiving other strata floating at the same elevation at which we were sailing, which from their appearance I judged to be highly charged with electricity, I considered it prudent to discharge twenty pounds of ballast, and we rose half a mile above our former elevation, where I considered we were perfectly safe and beyond their influence. I observed, amongst other phenomena, that at every discharge of thunder all the detached pillars of clouds within the distance of a mile around became attracted and appeared to concentrate their force towards the first body of clouds alluded to, leaving the atmosphere clear and calm beneath and around us.

"With very trifling variations we continued the same course until 7.15 p.m., when we descended to within 500 feet of the earth; but, perceiving from the disturbed surface of the rivers and lakes that a strong wind existed near the earth, we again ascended and continued our course till 7.30 p.m., when a final descent was safely effected in a meadow field in the parish of Crawley in Surrey, situated between Guildford and Horsham, and fifty-eight miles from Newbury. This stormy voyage was performed in one hour and a half."

It was after Green had followed his profession for fifteen years that he was called upon to undertake the management of an aerial venture, which, all things considered, has never been surpassed in genuine enterprise and daring. The conception of the project was due to Mr. Robert Hollond, and it took shape in this way. This gentleman, fresh from Cambridge, possessed of all the ardour of early manhood, as also of adequate means, had begun to devote himself with the true zeal of the enthusiast to the pursuit of ballooning, finding due opportunity for this in his friendship with Mr. Green, who enjoyed the management of the fine balloon made for ascents at the then popular Vauxhall Gardens. In the autumn of 1836 the proprietors of this balloon, contemplating making an exhibition of an ascent from Paris, and requiring their somewhat fragile property to be conveyed to that city, Mr. Hollond boldly came forward and offered to transfer it thither, and, as nearly as this might be possible, by passage through the sky. The proposal was accepted, and Mr. Holland, in conjunction with Green, set about the needful preparations. These, as will appear, were on an extraordinary scale, and no blame is to be imputed on that account, as a little consideration will show. For the venture proposed was not to be that of merely crossing the Channel, which, as we have seen, had been successfully effected no less than fifty years before. The voyage in contemplation was to be from London; it was, moreover, to be pursued through a long, moonless winter's night, and under conditions of which no living aeronaut had had actual experience.

Calculation, based on a sufficient knowledge of fast upper currents, told that their course, ere finished, might be one of almost indefinite length, and it is not too much to say that no one, with the knowledge of that day, could predict within a thousand miles where the dawn of the next day might find them. The equipment, therefore, was commensurate with the possible task before them. To begin with, they limited their number to three in all — Mr. Hollond, as chief and keeper of the log; Mr. Green, as aeronaut; and an enthusiastic colleague, Mr. Monck Mason, as the chronicler of the party. Next, they provided themselves with passports to all parts of the Continent; and then came the fitting out and victualling of the aerial craft itself, calculated to carry some 90,000 cubic feet of gas, and a counterpoise of a ton of ballast, which took the form partly of actual provisions in large quantity, partly of gear and apparatus, and for the rest of sand and also lime, of which more anon. Across the middle of the car was fixed a bench to serve as table, and also as a stage for the winding in and out of an enormous trail rope a thousand feet long, designed by Mr. Green to meet the special emergencies of the voyage. At the bottom of the car was spread a large cushion to serve the purposes of rest. When all was in readiness unfitness of weather baulked the travellers for some days, but Monday, the 7th of November, was judged a favourable day, so that the inflation was rapidly proceeded with, and at 1.30 p.m. the "Monstre Balloon," as it was entitled in the "Ingoldsby Legends," left the earth on her eventful and ever memorable voyage. The weather was fine and promising, and, rising with a moderate breeze from the N.W., they began to traverse the northern parts of Kent, while light, drifting upper clouds gave indication of other possible currents. Mr. Hollond was precise in the determination of times and of all readings and we learn that at exactly 2.48 p.m. they were crossing the Medway, six miles west of Rochester, while at 4.5 p.m. the lofty towers of Canterbury were well in view, two miles to the east, and here a little function was well carried out. Green had twice ascended from this city under patronage of the authorities, and the idea occurred to the party that it would be a graceful compliment to drop a message to the Mayor as they passed. A suitable note, therefore, quickly written, was dismissed in a parachute, and it may be mentioned that this, as also a similar missive addressed later to the Mayor of Dover, were duly received and acknowledged.

At a quarter past four they sighted the sea, and here, the air beginning to grow chill, the balloon dropped earthward, and for some miles they skimmed the ground, disturbing the partridges, scattering the rooks, and keeping up a running conversation the while with labourers and passers below. In this there was exercise of perfectly proper aerial seamanship, such as moreover presently led to an exhibition of true science. To save ballast is, with a balloon, to prolong life, and this may often best be done by flying low, which doubtless was Green's present intention. But soon his trained eye saw that the ground current which now carried them was leading them astray. They were trending to the northward, and so far out of their course that they would soon make the North Foreland, and so be carried out over the North Sea far from their desired direction. Thereupon Green attempted to put in practice his theory, already spoken of, of steering by upper currents, and the event proved his judgment peculiarly correct. "Nothing," wrote Mr. Monck Mason, "could exceed the beauty of the manoeuvre, to which the balloon at once responded, regaining her due course, and, in a matter of a few minutes only, bearing the voyagers almost vertically over the castle of Dover in the exact line for crossing the straits between that town and Calais."

So far all was well, and success had been extraordinary; but from this moment they became faced with new conditions, and with the grave trouble of uncertainty. Light was failing, the sea was before them, and—what else thenceforth? 4.48 p.m. was recorded as the moment when the first line of breaking waves was seen directly below them, and then the English coast line began rapidly to fade out from their view. But, ahead, the obscurity was yet more intense, for clouds, banked up like a solid wall, crowned along its frowning heights, with "parapets and turrets and batteries and bastions," and, plunging into this opposing barrier, they were quickly buried in blackness, losing at the same time over the sea all sound from earth soever. So for a short hour's space, when the sound of waves once again broke in upon them, and immediately afterwards emerging from the dense cloud (a sea-fog merely) they found themselves immediately over the brilliantly lighted town of Calais. Seeing this, the travellers attempted to signal by igniting and lowering a Bengal Light, which was directly followed by the beating of drums from below.

It adds a touch of reality, as well as cheerfulness, to the narrative to read that at this period of their long journey the travellers apply themselves to a fair, square meal, the first for twelve hours, despite the day's excitement and toil. We have an entry among the stores of the balloon of wine bottles and spirit flasks, but there is no mention of these being requisitioned at this period. The demand seems rather to have been for coffee—coffee hot; and this by a novel device was soon prepared. It goes without saying that a fire or flame of any kind, except with special precautions, is inadmissable in a balloon; but a cooking heat, sufficient for the present purpose, was supplied from the store of lime, a portion of which, being placed in a suitably contrived vessel and slaked quickly, procured the desired beverage.

This meal now indulged in seems to have been heartily and happily enjoyed; and from this point, for a while, the narrative becomes that of enthusiastic and delighted travellers. In the gloom below, for leagues around, they regarded the scattered fires of a watchful population, with here and there the lights of larger towns, and the contemplation begot romantic reveries. "Were they not amid the vast solitudes of the skies, in the dead of night, unknown and unnoticed, secretly and silently reviewing kingdoms, exploring territories, and surveying cities all clothed in the dark mantle of mystery?" Presently they identified the blazing city of Liege, with the lurid lights of extensive outlying iron works, and this was the last visible sign they caught of earth that night; save, at least, when occasional glimpses of lightning momentarily and dimly outlined the world in the abyss below.

Ere long, they met with their first discomfort, which they seem to have regarded as a most serious one, namely, the accidental dropping overboard of their cherished coffee-boiling apparatus. With its loss their store of lime became useless, save as ballast, and for this it was forthwith utilised until nothing remained but the empty lime barrel itself, which, being regarded as an objectionable encumbrance, it was desirable to get rid of, were it not for the risk involved in rudely dropping it to earth. But the difficulty was met. They possessed a suitable small parachute, and, attached to this, the barrel was allowed to float earthward.

As hours advanced, the blackness of night increased, and their impressions appear somewhat strange to anyone familiar with ordinary night travel in the sky. Mr. Monck Mason compares their progress through the darkness to "cleaving their way through an interminable mass of black marble." Then, presently, an unaccountable object puzzles and absorbs the attention of all the party for a long period. They were gazing open-mouthed at a long narrow avenue of feeble light, which, though apparently belonging to earth, was too long and regular for a river, and too broad for a canal or road, and it was only after many futile imaginings that they discovered they were simply looking at a stay rope of the balloon hanging far out over the side.

Somewhat later still, there was a more serious claim upon the imagination. It was half-past three in the morning, and the balloon, which, to escape from too low an altitude, had been liberally lightened, had now at high speed mounted to a vast height. And then, amid the black darkness and dead silence of that appalling region, suddenly overhead came the sound of an explosion, followed by the violent rustling of the silk, while the car jerked violently, as though suddenly detached from its hold. This was the idea, leading to the belief that the balloon had suddenly exploded, and that they were falling headlong to earth. Their suspense, however, cannot have been long, and the incident was intelligible enough, being due to the sudden yielding of stiffened net and silk under rapid expansion caused by their speedy and lofty ascent.

The chief incidents of the night were now over, until the dawn arrived and began to reveal a strange land, with large tracts of snow, giving place, as the light strengthened, to vast forests. To their minds these suggested the plains of Poland, if not the steppes of Russia, and, fearing that the country further forward might prove more inhospitable, they decided to come to earth as speedily as possible. This, in spite of difficult landing, they effected about the hour that the waking population were moving abroad, and then, and not till then, they learned the land of their haven—the heart of the German forests. Five hundred miles had been covered in eighteen hours from start to finish!

CHAPTER VII. CHARLES GREEN — FURTHER ADVENTURES.

All history is liable to repeat itself, and that of aeronautics forms no exception to the rule. The second year after the invention of the balloon the famous M. Blanchard, ascending from Frankfort, landed near Weilburg, and, in commemoration of the event, the flag he bore was deposited among the archives in the ducal palace of that town. Fifty-one years passed by when, outside the same city, a yet more famous balloon effected its landing, and with due ceremony its flag is presently laid beside that of Blanchard in the same ducal palace. The balloon of the "Immortal Three," whose splendid voyage has just been recounted, will ever be known by the title of the Great Nassau Balloon, but the neighbourhood of its landing was that of the town of Weilburg, in the Duchy of Nassau, whither the party betook themselves, and where, during many days, they were entertained with extravagant hospitality and honour until business recalled Mr. Hollond home.

Green had now made upwards of two hundred ascents, and, though he lived to make a thousand, it was impossible that he could ever eclipse this last record. It is true that the same Nassau balloon, under his guidance, made many other most memorable voyages, some of which it will be necessary to dwell on. But, to preserve a better chronology, we must first, without further digression, approach an event which fills a dark page in our annals; and, in so doing, we have to transfer our attention from the balloon itself to its accessory, the parachute.

Twenty-three years before our present date, that is to say in 1814, Mr. Cocking delivered his views as to the proper form of the parachute before the Society of Arts, who, as a mark of approval, awarded him a medal. This parachute, however, having never taken practical shape, and only existing, figuratively speaking, in the clouds, seemed unlikely to find its way there in reality until the success of the Nassau adventure stirred its inventor to strenuous efforts to give it an actual trial. Thus it came about that he obtained Mr. Green's co-operation in the attempt he now undertook, and, though this ended disastrously, for Mr. Cocking, the great professional aeronaut can in no way soever be blamed for the tragic event.

The date of the trial was in July, 1837. Mr. Cocking's parachute was totally different in principle from that form which, as we have seen, had met with a fair measure of success at the hands of early experimenters; and on the eve of its trial it was strongly denounced and condemned in the London Press by the critic whom we have recently so freely quoted, Mr. Monck Mason.

This able reasoner and aeronaut pointed out that the contrivance about to be tested aimed at obviating two principal drawbacks which the parachute had up to that time presented, namely (1) the length of time which elapses before it becomes sufficiently expanded, and (2) the oscillatory movement which accompanies the descent. In this new endeavour the inventor caused his machine to be fixed rigidly open, and to assume the shape of an inverted cone. In other words, instead of its being like an umbrella opened, it rather resembled an umbrella blown inside out. Taking, then, the shape and dimensions of Mr. Cocking's structure as a basis for mathematical calculation, as also its weight, which for required strength he put at 500 lbs. Mr. Monck Mason estimated that the adventurer and his machine must attain in falling a velocity of some twelve miles an hour. In fact, his positive prediction was that one of two events must inevitably take place. "Either the parachute would come to the ground with a force incompatible with the safety of the individual, or should it be attempted to make it sufficiently light to resist this conclusion, it must give way beneath the forces which will develop in the descent."

This emphatic word of warning was neglected, and the result of the terrible experiment can best be gathered from two principal sources. First, that of a special reporter writing from terra-firma, and, secondly, that of Mr. Green himself, who gives his own observations as made from the balloon in which he took the unfortunate man and his invention into the sky.

The journalist, who first speaks of the enormous concourse that gathered to see the ascent, not only within Vauxhall Gardens, but on every vantage ground without, proceeds to tell of his interview with Mr. Cocking himself, who, when questioned as to the danger involved, remarked that none existed for him, and that the greatest peril, if any, would attend the balloon when suddenly relieved of his weight. The proprietors of the Gardens, as the hour approached, did their best to dissuade the over-confident inventor, offering, themselves, to take the consequences of any public disappointment. This was again without avail, and so, towards 6 p.m., Mr. Green, accompanied by Mr. Spencer, a solicitor of whom this history will have more to tell, entered the balloon, which was then let up about 40 feet that the parachute might be affixed below. A little later, Mr. Cocking, casting aside his heavy coat and tossing off a glass of wine, entered his car and, amid deafening acclamations, with the band playing the National Anthem, the balloon and aeronauts above, and he himself in his parachute swinging below, mounted into the heavens, passing presently, in the gathering dusk, out of view of the Gardens.

The sequel should be gathered from Mr. Green's own narrative. Previous to starting, 650 lbs. of ballast had to be discarded to gain buoyancy sufficient to raise the massive machine. This, together with another 100 lbs., which was also required to be ejected owing to the cooling of the air, was passed out through a canvas tube leading downwards through a hole in the parachute, an ingenious contrivance which would prevent the sand thrown out from the balloon falling on the slender structure itself. On quitting the earth, however, this latter set up such violent oscillations that the canvas tube was torn away, and then it became the troublesome task of the aeronauts to make up their ballast into little parcels, and, as occasion required, to throw these into space clear of the swinging parachute below.

Despite all efforts, however, it was soon evident that the cumbersome nature of the huge parachute would prevent its being carried up quite so high as the inventor desired. Mr. Cocking had stipulated for an elevation of 7,000 feet, and, as things were, only 5,000 feet could be reached, at any rate, before darkness set in. This fact was communicated to Mr. Cocking, who promptly intimated his intention of leaving, only requesting to know whereabouts he was, to which query Mr. Spencer replied that they were on a level with Greenwich. The brief colloquy that ensued is thus given by Mr. Green: —

"I asked him if he felt quite comfortable, and if the practical trial bore out his calculation. Mr. Cocking replied, 'Yes, I never felt more comfortable or more delighted in my life,' presently adding, 'Well, now I think I shall leave you.' I answered, 'I wish you a very "Good Night!" and a safe descent if you are determined to make it and not use the tackle' (a contrivance for enabling him to retreat up into the balloon if he desired). Mr. Cocking's only reply was, 'Good-night, Spencer; Good-night, Green!' Mr. Cocking then pulled the rope that was to liberate himself, but too feebly, and a moment afterwards more violently, and in an instant the balloon shot upwards with the velocity of a sky rocket. The effect upon us at this moment was almost beyond description. The immense machine which suspended us between heaven and earth, whilst it appeared to be forced upwards with terrific violence and rapidity through unknown and untravelled regions amidst the howlings of a fearful hurricane, rolled about as though revelling in a freedom for which it had long struggled, but of which until that moment it had been kept in utter ignorance. It, at length, as if somewhat fatigued by its exertions, gradually assumed the motions of a snake working its way with extraordinary speed towards a given object. During this frightful operation the gas was rushing in torrents from the upper and lower valve, but more particularly from the latter, as the density of the atmosphere through which we were forcing our progress pressed so heavily on the valve at the top of the balloon as to admit of but a comparatively small escape by this aperture. At this juncture, had it not been for the application to our mouths of two pipes leading into an air bag, with which we had furnished ourselves previous to starting, we must within a minute have been suffocated, and so, but by different means, have shared the melancholy fate of our friend. This bag was formed of silk, sufficiently capacious to contain 100 gallons of atmospheric air. Prior to our ascent, the bag was inflated with the assistance of a pair of bellows with fifty gallons of air, so allowing for any expansion which might be produced in the upper regions. Into the end of this bag were introduced two flexible tubes, and the moment we felt ourselves to be going up in the manner just described, Mr.

Spencer, as well as myself, placed either of them in our mouths. By this simple contrivance we preserved ourselves from instantaneous suffocation, a result which must have ensued from the apparently endless volume of gas with which the car was enveloped. The gas, notwithstanding all our precautions, from the violence of its operation on the human frame, almost immediately deprived us of sight, and we were both, as far as our visionary powers were concerned, in a state of total darkness for four or five minutes."

Messrs. Green and Spencer eventually reached earth in safety near Maidstone, knowing nothing of the fate of their late companion. But of this we are sufficiently informed through a Mr. R. Underwood, who was on horseback near Blackheath and watching the aeronauts at the moment when the parachute was separated from the balloon. He noticed that the former descended with the utmost rapidity, at the same time swaying fearfully from side to side, until the basket and its occupant, actually parting from the parachute, fell together to earth through several hundred feet and were dashed to pieces.

It would appear that the liberation of the parachute from below the balloon had been carried out without hitch; indeed, all so far had worked well, and the wind at the time was but a gentle breeze. The misadventure, therefore, must be entirely attributed to the faulty manner in which the parachute was constructed. There could, of course, be only one issue to the sheer drop from such a height, which became the unfortunate Mr. Cocking's fate, but the very interesting question will have to be discussed as to the chances in favour of the aeronaut who, within his wicker car, while still duly attached to the balloon, may meet with a precipitate descent.

We may here fitly mention an early perilous experience of Mr. Green, due simply to the malice of someone never discovered. It appears that while Green's balloon, previous to an ascent, was on the ground, the cords attaching the car had been partly severed in such a way as to escape detection. So that as soon as the balloon rose the car commenced breaking away, and its occupants, Mr. Green and Mr. Griffiths, had to clutch at the ring, to which with difficulty they continued to cling. Meanwhile, the car remaining suspended by one cord only, the balloon was caused to hang awry, with the result that its upper netting began giving way, allowing the balloon proper gradually to escape through the bursting meshes, thus threatening the distracted voyagers with terrible disaster. The disaster, in fact, actually came to pass ere the party completed their descent, "the balloon, rushing through the opening in the net-work with a tremendous explosion, and the two passengers clinging to the rest of the gear, falling through a height said to be near a hundred feet. Both, though only with much time and difficulty, recovered from the shock."

In 1840, three years after the tragic adventure connected with Mr. Cocking's parachute trial, we find Charles Green giving his views as to the practicability of carrying out a ballooning enterprise which should far excel all others that had hitherto been attempted. This was nothing less than the crossing of the Atlantic from America to England. There is no shadow of doubt that the adventurous aeronaut was wholly in earnest in the readiness he expressed to embark on the undertaking should adequate funds be forthcoming; and he discusses the possibilities with singular clearness and candour. He maintains that the actual difficulties resolve themselves into two only: first, the maintenance of the balloon in the sky for the requisite period of time; and, secondly, the adequate control of its direction in space. With respect to the first difficulty, he points out the fact to which we have already referred, namely, that it is impossible to avoid the fluctuations of level in a balloon's course, "by which it constantly becomes alternately subjected to escape of gas by expansion, and consequent loss of ballast, to furnish an equivalent diminution of weight." Taking his own balloon of 80,000 cubic feet by way of example, he shows that this, fully inflated on the earth, would lose 8,000 cubic feet of gas by expansion in ascending only 3,000 feet. Moreover, the approach of night or passage through cloud or falling rain would occasion chilling of the gas or accumulation of moisture on the silk, in either case necessitating the loss of ballast, the store of which is always the true measure of the balloon's life.

To combat the above difficulty Green sanguinely relies on his favourite device of a trail or guide rope, whose function, being that of relieving the balloon of a material weight as it approaches the earth, could, he supposed, be made to act yet more efficiently when over the sea in the following manner. Its length, suspended from the ring, being not less than 2,000 feet, it should have attached at its lower end at certain intervals a number of small, stout waterproof canvas bags, the apertures of which should be contrived to admit water, but to oppose its return. Between these bags were to be conical floats, to support any length of the rope that might descend on the sea. Now, should the balloon commence descending, it would simply deposit a certain portion of rope on the water until it regained equilibrium at no great decrease of altitude, and would thus continue its course until alteration of conditions should cause it to recommence rising, when the weight of water now collected in the bags would play its part in preventing the balloon from soaring up into space. With such a contrivance Green allowed himself to imagine that he could keep a properly made balloon at practically the same altitude for a period of three months if required.

The difficulty of maintaining a due course was next discussed, and somewhat speedily disposed of. Here Green relied on the results of his own observation, gathered during 275 ascents, and stated his conviction that there prevails a uniformity of upper wind currents that would enable him to carry out his bold projects successfully. His contention is best given in his own words:

"Under whatever circumstances," he says, "I made my ascent, however contrary the direction of the winds below, I uniformly found that at a certain elevation, varying occasionally, but always within 10,000 feet of the earth, a current from the west or rather from the north of west, invariably travailed, nor do I recollect a single instance in which a different result ensued." Green's complete scheme is now sufficiently evident. He was to cross the Atlantic practically by the sole assistance of upper currents and his guide rope, but on this latter expedient, should adverse conditions prevail, he yet further relied, for he conceived that the rope could have attached to its floating end a water drag, which would hold the balloon in check until favouring gales returned.

Funds, apparently, were not forthcoming to allow of Mr. Green's putting his bold method to the test; but we find him still adhering with so much zeal to his project that, five years later, he made, though again unsuccessfully, a second proposal to cross the Atlantic by balloon. He still continued to make many and most enterprising ascents, and one of a specially sensational nature must be briefly mentioned before we pass on to regard the exploits of other aeronauts.

It was in 1841 on the occasion of a fete at Cremorne House, when Mr. Green, using his famous Nassau balloon, ascended with a Mr. Macdonnell. The wind was blowing with such extreme violence that Rainham, in Essex, about twenty miles distant, was reached in little more than a quarter of an hour, and here, on nearing the earth, the grapnel, finding good hold, gave a wrench to the balloon that broke the ring and jerked the car completely upside down, the aeronauts only escaping precipitation by holding hard to the ropes. A terrific steeplechase ensued, in which the travellers were dragged through stout fencing and other obstacles till the balloon, fairly emptied of gas, finally came to rest, but not until some severe injuries had been received.

CHAPTER VIII. JOHN WISE — THE AMERICAN AERONAUT.

By this period the domination of the air was being pursued in a fresh part of the world. England and her Continental neighbours had vied with each in adding to the roll of conquests, and it could hardly other be supposed that America would stand by without taking part in the campaign which was now being revived with so much fresh energy in the skies.

The American champion who stepped forward was Mr. John Wise, of Lancaster, Pa., whose career, commencing in the year 1835, we must now for a while follow. Few attempts at ballooning of any kind had up to that time been made in all America. There is a record that in December, 1783, Messrs. Rittenhouse and Hopkins, Members of the Philosophical Academy of Philadelphia, instituted experiments with an aerial machine consisting of a cage to which forty-seven small balloons were harnessed. In this strange craft a carpenter, by name Wilcox, was induced to ascend, which, it is said, he did successfully, remaining in the air for ten minutes, when, finding himself near a river, he sought to come to earth again by opening several of his balloons. This brought about an awkward descent, attended, however, by no more serious accident than a dislocated wrist. Mr. Wise, on the other hand, states that Blanchard had won the distinction of making the first ascent in the New World in 1793 in Philadelphia on which occasion Washington was a spectator; and a few years afterwards other Frenchmen gave exhibitions, which, however, led to no real development of the new art on this, the further side of the Atlantic. Thus the endeavours we are about to describe were those of an independent and, at the same time, highly, practical experimentalist, and on this account have a special value of their own.

The records that Wise has left of his investigations begin at the earliest stage, and possess the charm of an obvious and somewhat quaint reality. They commence with certain crude calculations which would seem to place no limit to the capabilities of a balloon. Thus, he points out that one of "the very moderate size of 400 feet diameter" would convey 13,000 men. "No wonder, then," he continues, "the citizens of London became alarmed during the French War, when they mistook the appearance of a vast flock of birds coming towards the Metropolis for Napoleon's army apparently coming down upon them with this new contrivance."

Proceeding to practical measures, Wise's first care was to procure some proper material of which to build an experimental balloon of sufficient size to lift and convey himself alone. For this he chose ordinary long-cloth, rendered gas-tight by coats of suitable varnish, the preparation of which became with him, as, indeed, it remains to this day, a problem of chief importance and difficulty. Perhaps it hardly needs pointing out that the varnish of a balloon must not only be sufficiently elastic not to crack or scale off with folding or unavoidable rough usage, but it must also be of a nature to resist the common tendency of such substances to become adherent or "tacky." Wise determined on bird lime thinned with linseed oil and ordinary driers. With this preparation he coated his material several times both before and after the making up, and having procured a net, of which he speaks with pride, and a primitive sort of car, of which he bitterly complains, he thought himself sufficiently equipped to embark on an actual ascent, which he found a task of much greater practical difficulty than the mere manufacture of his air ship. For the inflation by hydrogen of so small a balloon as his was he made more than ample provision in procuring no less than fifteen casks of 130 gallons capacity each. He also duly secured a suitable filling ground at the corner of Ninth and Green Streets, Philadelphia, but he made a miscalculation as to the time the inflation would demand, and this led to unforeseen complications, for as yet he knew not the way of a crowd which comes to witness a balloon ascent.

Having all things in readiness, and prudently waiting for fair weather, he embarked on his grand experiment on the 2nd of May, 1835, announcing 4 p.m. as the hour of departure. But by that time the inflation, having only proceeded for three hours, the balloon was but half full, and then the populace began to behave as in such circumstances they always will. They were incredulous, and presently grew troublesome. In vain the harnessing of the car was proceeded with as though all were well. For all was not well, and when the aeronaut stepped into his car with only fifteen pounds of sand and a few instruments he must have done so with much misgiving. Still, he had friends around who might have been useful had they been less eager to help. But these simply crowded round him, giving him no elbow room, nor opportunity for trying the "lift" of his all-too-empty globe. Moreover, some would endeavour to throw the machine upward, while others as strenuously strove to keep it down, and at last the former party prevailed, and the balloon, being fairly cast into the air, grazed a neighbouring chimney and then plunged into an adjacent plot, not, however, before the distracted traveller had flung away all his little stock of sand. There now was brief opportunity for free action, and to the first bystander who came running up Wise gave the task of holding the car in check. To the next he handed out his instruments, his coat, and also his boots, hoping thus to get away; but his chance had not yet come, for once again the crowd swarmed round him, keeping him prisoner with good-natured but mistaken interference, and drowning his voice with excited shouting. Somehow, by word and gesture, he gave his persecutors to understand that he wished to speak, and then he begged them only to give him a chance, whereupon the crowd fell back, forming a ring, and leaving only one man holding the car. It was a moment of suspense, for Wise calculated that he had only parted with some eighteen pounds since his first ineffectual start from the filling ground; but it was enough, and in another moment he was sailing up clear above the crowd. So great, as has been already shewn, is often the effect of parting with the last few pounds of dead weight in a well-balanced balloon.

Such was the first "send off" of the future great balloonist, destined to become the pioneer in aeronautics on the far side of the Atlantic. The balloon ascended to upwards of a mile, floating gradually away, but at its highest point it reached a conflict of currents, causing eddies from which Wise escaped by a slight decrease of weight, effected by merely cutting away the wreaths of flowers that were tied about his car. A further small substitute for ballast he extemporised in the metal tube inserted in the neck of his fabric, and this he cast out when over the breadth of the Delaware, and he describes it as falling with a rustling sound, and striking the water with a splash plainly heard at more than a mile in the sky. After an hour and a quarter the balloon spontaneously and steadily settled to earth.

An ascent carried out later in the same summer led to a mishap, which taught the young aeronaut an all-important lesson. Using the same balloon and the same mode of inflation, he got safely and satisfactorily away from his station in the town of Lebanon, Pa., and soon found himself over a toll gate in the open country, where the gate keeper in banter called up to him for his due. To this summons Wise, with heedless alacrity, responded in a manner which might well have cost him dear. He threw out a bag of sand to represent his toll, and, though he estimated this at only six pounds, it so greatly accelerated his ascent that he shortly found himself at a greater altitude than he ever after attained. He passed through mist into upper sunshine, where he experienced extreme cold and ear-ache, at which time, seeking the natural escape from such trouble, he found to his dismay that the valve rope was out of reach. Thus he was compelled to allow the balloon to ascend yet higher, at its own will; and then a terrible event happened.

By mischance the neck of his balloon, which should have been open, was out of reach and folded inwards in such a way as to prevent the free escape of the gas, which, at this great altitude, struggled for egress with a loud humming noise, giving him apprehensions of an accident which very shortly occurred, namely, the bursting of the lower part of his balloon with a loud report. It happened, however, that no extreme loss of gas ensued, and he commenced descending with a speed which, though considerable, was not very excessive. Still, he was eager to alight in safety, until a chance occurrence made him a second time that afternoon guilty of an act of boyish impetuosity. A party of volunteers firing a salute in his honour as he neared the ground, he instantly flung out papers, ballast, anything he could lay his hands on, and once again soared to a great height with his damaged balloon. He could then do no more, and presently subsiding to earth again, he acquired the welcome knowledge that even in such precarious circumstances a balloon may make a long fall with safety to its freight.

Mr. Wise's zeal and indomitable spirit of enterprise led to speedy developments of the art which he had espoused; the road to success being frequently pointed out by failure or mishap. He quickly discarded the linen balloon for one of silk on which he tried a new varnish composed of linseed oil and india-rubber, and, dressing several gores with this, he rolled them up and left them through a night in a drying loft, with the result that the next day they were disintegrated and on the point of bursting into flame by spontaneous combustion. Fresh silk and other varnish were then tried, but with indifferent success. Next he endeavoured to dispense with sewing, and united the gores of yet another balloon by the mere adhesiveness of the varnish and application of a hot iron. This led to a gaping seam developing at the moment of an ascent, and then there followed a hasty and hazardous descent on a house-top and an exciting rescue by a gentleman who appeared opportunely at a third storey window. Further, another balloon had been destroyed, and Wise badly burned, at a descent, owing to a naked light having been brought near the escaping gas. It is then without wonder that we find him after this temporarily bankrupt, and resorting to his skill in instrument-making to recover his fortunes. Only, however, for a few months, after which he is before the public once more as a professional aeronaut. He now adopts coal gas for inflation, and incidents of an impressive nature crowd into his career, forcing important facts upon him. The special characteristics of his own country present peculiar difficulties; broad rivers and vast forests become serious obstacles. He is caught in the embrace of a whirlwind; he narrowly escapes falling into a forest fire; he is precipitated, but harmlessly, into a pine wood. Among other experiments, he makes a small copy of Mr. Cocking's parachute, and drops it to earth with a cat as passenger, proving thereby that that unfortunate gentleman's principle was really less in fault than the actual slenderness of the material used in his machine.

We now approach one of Wise's boldest, and at the same time most valuable, experiments. It was the summer of 1839, and once again the old trouble of spontaneous combustion had destroyed a silk balloon which was to have ascended at Easton, Pa. Undeterred, however, Wise resolutely advertised a fresh attempt, and, with only a clear month before the engagement, determined on hastily rigging up a cambric muslin balloon, soaking it in linseed oil and essaying the best exhibition that this improvised experiment could afford. It was intended to become a memorable one, inasmuch as, should he meet with no hindrance, his determination was nothing less than that of bursting this balloon at a great height, having firmly convinced himself that the machine in these circumstances would form itself into a natural parachute, and bring him to earth with every chance in favour of safety. In his own words, "Scientific calculations were on his side with a certainty as great and principles as comprehensive as that a pocket-handkerchief will not fall as rapidly to the ground when thrown out of a third storey window as will a brick."

His balloon was specially contrived for the experiment in hand, having cords sewn to the upper parts of its seams, and then led down through the neck, where they were secured within reach, their office being that of rending the whole head of the balloon should this be desired. On this occasion a cat and a dog were taken up, one of these being let fall from a height of 2,000 feet in a Cocking's parachute, and landing in safety, the other being similarly dismissed at an altitude of 4,000 feet in an oiled silk balloon made in the form of a collapsed balloon, which, after falling a little distance, expanded sufficiently to allow of its descending with a safe though somewhat vibratory motion. Its behaviour, at any rate, fully determined Wise on carrying out his own experiment.

Being constructed entirely for the main object in view, the balloon had no true opening in the neck beyond an orifice of about an inch, and by the time a height of 13,000 feet had been reached the gas was streaming violently through this small hole, the entire globe being expanded nearly to bursting point, and the cords designed for rending the balloon very tense. At this critical period Wise owns to having experienced considerable nervous excitement, and observing far down a thunderstorm in progress he began to waver in his mind, and inclined towards relieving the balloon of its strain, and so abandoning his experiment, at least for the present. He remembers pulling out his watch to make a note of the hour, and, while thus occupied, the straining cords, growing tenser every moment, suddenly took charge of the experiment and burst the balloon of their own accord. The gas now rushed from the huge rent above tumultuously and in some ten seconds had entirely escaped, causing the balloon to descend rapidly, until the lower part of the muslin, doubling in upwards, formed a species of parachute after the manner intended. The balloon now came down with zig-zag descent, and finally the car, striking the earth obliquely, tossed its occupant out into a field unharmed. Shortly after this Wise experimented with further success with an exploded balloon.

It is not a little remarkable that this pioneer of aeronautics in American—a contemporary of Charles Green in England, but working and investigating single-handed on perfectly independent lines—should have arrived at the same conclusions as did Green himself as to the possibility, which, in his opinion, amounted to a certainty, of being able to cross the Atlantic by balloon if only adequate funds were forth-coming. So intent was he on his bold scheme that, in the summer of 1843, he handed to the Lancaster Intelligencer a proclamation, which he desired might be conveyed to all publishers of newspapers on the globe. It contained, among other clauses, the following:—

"Having from a long experience in aeronautics been convinced that a constant and regular current of air is blowing at all times from west to east, with a velocity of from twenty to forty and even sixty miles per hour, according to its height from the earth, and having discovered a composition which renders silk or muslin impervious to hydrogen gas, so that a balloon may be kept afloat for many weeks, I feel confident with these advantages that a trip across the Atlantic will not be attended with as much real danger as by the common mode of transition. The balloon is to be 100 feet in diameter, giving it a net ascending power of 25,000 lbs." It was further stated that the crew would consist of three persons, including a sea navigator, and a scientific landsman. The specifications for the transatlantic vessel were also to include a seaworthy boat in place of the ordinary car. The sum requisite for this enterprise was, at the time, not realised; but it should be mentioned that several years later a sufficient sum of money was actually subscribed. In the summer of 1873 the proprietors of the New York Daily Graphic provided for the construction of a balloon of no less than 400,000 cubic feet capacity, and calculated to lift 14,000 lbs. It was, however, made of bad material; and, becoming torn in inflation, Wise condemned and declined to use it. A few months later, when it had been repaired, one Donaldson and two other adventurers, attempting a voyage with this ill-formed monster, ascended from New York, and were fortunate in coming down safely, though not without peril, somewhere in Connecticut.

Failing in his grand endeavour, Wise continued to follow the career of a professional aeronaut for some years longer, of which he has left a full record, terminating with the spring of 1848. His ascents were always marked by carefulness of detail, and a coolness and courage in trying circumstances that secured him uniform success and universal regard. He was, moreover, always a close and intelligent observer, and many of his memoranda are of scientific value.

His description of an encounter with a storm-cloud in the June of 1843 has an interest of its own, and may not be considered overdrawn. It was an ascent from Carlisle, Pa., to celebrate the anniversary of Bunker's Hill, and Wise was anxious to gratify the large concourse of people assembled, and thus was tempted, soon after leaving the ground, to dive up into a huge black cloud of peculiarly forbidding aspect. This cloud appeared to remain stationary while he swept beneath it, and, having reached its central position, he observed that its under surface was concave towards the earth, and at that moment he became swept upwards in a vortex that set his balloon spinning and swinging violently, while he himself was afflicted with violent nausea and a feeling of suffocation. The cold experienced now became intense, and the cordage became glazed with ice, yet this had no effect in checking the upward whirling of the balloon. Sunshine was beyond the upper limits of the cloud; but this was no sooner reached than the balloon, escaping from the uprush, plunged down several hundred feet, only to be whirled up again, and this recip-rocal motion was repeated eight or ten times during an interval of twenty minutes, in all of which time no expenditure of gas or discharge of ballast enabled the aeronaut to regain any control over his vessel.

Statements concerning a thunderstorm witnessed at short range by Wise will compare with other accounts. The thunder "rattled" without any reverberations, and when the storm was passing, and some dense clouds moving in the upper currents, the "surface of the lower stratum swelled up suddenly like a boiling cauldron, which was immediately followed by the most brilliant ebullition of sparkling coruscations." Green, in his stormy ascent from Newbury, England, witnessed a thunderstorm below him, as will be remembered, while an upper cloud stratum lay at his own level. It was then that Green observed that "at every discharge of thunder all the detached pillars of clouds within the distance of a mile around became attracted."

The author will have occasion, in due place, to give personal experiences of an encounter with a thunderstorm which will compare with the foregoing description.

CHAPTER IX. EARLY METHODS AND IDEAS.

Before proceeding to introduce the chief actors and their achievements in the period next before us, it will be instructive to glance at some of the principal ideas and methods in favour with aeronauts up to the date now reached. It will be seen that Wise in America, contrary to the practice of Green in our own country, had a strong attachment to the antique mode of inflation with hydrogen prepared by the vitriolic process; and his balloons were specially made and varnished for the use of this gas. The advantage which he thus bought at the expense of much trouble and the providing of cumbersome equipment was obvious enough, and may be well expressed by a formula which holds good to-day, namely, that whereas 1,000 cubic feet of hydrogen is capable of lifting 7 lbs., the same quantity of coal gas of ordinary quality will raise but 35 lbs. The lighter gas came into all Wise's calculations for bolder schemes. Thus, when he discusses the possibility of using a metal balloon, his figures work out as follows: If a balloon of 200 feet diameter were constructed out of copper, weighing one pound to the square foot; if, moreover, some six tons were allowed for the weight of car and fastenings, an available lifting power would remain capable of raising 45 tons to an altitude of two miles. This calculation may appear somewhat startling, yet it is not only substantially correct, but Wise entertained no doubt as to the practicability of such a machine. For its inflation he suggests inserting a muslin balloon filled with air within the copper globe, and then passing hydrogen gas between the muslin and copper surfaces, which would exclude the inner balloon as the copper one filled up.

His method of preparing hydrogen was practically that still adopted in the field, and seems in his hands to have been seldom attended with difficulty. With eight common 130-gallon rum puncheons he could reckon on evolving 5,000 cubic feet of gas in an hour, using his elements in the following proportions: water, 560 lbs.; sulphuric acid (sp. g. 1.85), 144 lbs.; iron turnings, 125 lbs. The gas, as given off, was cooled and purified by being passed through a head of water kept cool and containing lime in solution. Contrasted with this, we find it estimated, according to the practice of this time, that a ton of good bituminous coal should yield 10,000 cubic feet of carburetted hydrogen fit for lighting purposes, and a further quantity which, though useless as an illuminant, is still of excellent quality for the aeronaut.

It would even seem from a statement of Mr. Monck Mason that the value of coke in his day largely compensated for the cost of producing coal gas, so that in a large number of Green's ascents no charge whatever was made for gas by the companies that supplied him.

Some, at least, of the methods formerly recommended for the management of free balloons must in these days be modified. Green, as we have seen, was in favour of a trail rope of inordinate length, which he recommended both as an aid to steering and for a saving of ballast. In special circumstances, and more particularly over the sea, this may be reckoned a serviceable adjunct, but over land its use, in this country at least, would be open to serious objection. The writer has seen the consternation, not to say havoc, that a trail rope may occasion when crossing a town, or even private grounds, and the actual damage done to a garden of hops, or to telegraph or telephone wires, may be very serious indeed. Moreover, the statement made by some early practitioners that a trail rope will not catch so as to hold fast in a wood or the like, is not to be relied on, for an instance could be mentioned coming under the writer's knowledge where such a rope was the source of so much trouble in a high wind that it had to be cut away.

The trouble arose in this way. The rope dragged harmlessly enough along the open ground. It would, likewise, negotiate exceedingly well a single tree or a whole plantation, catching and releasing itself with only such moderate tugs at the car as were not disturbing; but, presently, its end, which had been caught and again released by one tree, swung free in air through a considerable gap to another tree, where, striking a horizontal bough, it coiled itself several times around, and thus held the balloon fast, which now, with the strength of the wind, was borne to the earth again and again, rebounding high in air after each impact, until freedom was gained only by the sacrifice of a portion of the rope.

Wise recommends a pendant line of 600 or 800 feet, capable of bearing a strain of 100 lbs., and with characteristic ingenuity suggests a special use which can be made of it, namely, that of having light ribbons tied on at every hundred feet, by means of which the drifts of lower currents may be detected. In this suggestion there is, indeed, a great deal of sound sense; for there is, as will be shown hereafter, very much value to be attached to a knowledge of those air rivers that are flowing, often wholly unsuspected, at various heights. Small parachutes, crumpled paper, and other such-like bodies as are commonly thrown out and relied on to declare the lower drifts, are not wholly trustworthy, for this reason—that air-streams are often very slender, mere filaments, as they are sometimes called, and these, though setting in some definite direction, and capable of entrapping and wafting away some small body which may come within their influence, may not affect the travel of so big an object as a balloon, which can only partake of some more general air movement.

Wise, by his expedient of tying ribbons at different points to his trail rope, would obtain much more correct and constant information respecting those general streams through which the pendant rope was moving. A similar expedient adopted by the same ingenious aeronaut is worthy of imitation, namely, that of tying ribbons on to a rod projecting laterally from the car. These form a handy and constant telltale as to the flight of the balloon, for should they be fluttering upwards the sky sailor at once knows that his craft is descending, and that he must act accordingly.

The material, pure silk, which was universally adopted up to and after the period we are now regarding, is not on every account to be reckoned the most desirable. In the first place, its cost alone is prohibitive, and next, although lighter than any kind of linen, strength for strength, it requires a greater weight of varnish, which, moreover, it does not take so kindly as does fabric made of vegetable tissue. Further, paradoxical as it may appear, its great strength is not entirely an advantage. There are occasions which must come into the experience of every zealous aeronaut when his balloon has descended in a rough wind, and in awkward country. This may, indeed, happen even when the ascent has been made in calm. Squalls of wind may spring up at short notice, or after traversing only two or three counties a strong gale may be found on the earth, though such was absent in the starting ground. This is more particularly the case when the landing chances to be on high ground in the neighbourhood of the sea. In these circumstances, the careful balloonist, who will generally be forewarned by the ruffle on any water he may pass, or by the drift of smoke, the tossing of trees, or by their very rustling or "singing" wafted upwards to him, will, if possible, seek for his landing place the lee of a wood or some other sheltered spot. But, even with all his care, he will sometimes find himself, on reaching earth, being dragged violently across country on a mad course which the anchor cannot check. Now, the country through which he is making an unwilling steeplechase may be difficult, or even dangerous. Rivers, railway cuttings, or other undesirable obstacles may lie ahead, or, worse yet, such a death trap as in such circumstances almost any part of Derbyshire affords, with its stone walls, its precipitous cliffs, and deep rocky dells. To be dragged at the speed of an express train through territory of this description will presently mean damage to something, perhaps to telegraph poles, to roofs, or crops, and if not, then to the balloon itself. Something appertaining to it must be victimised, and it is in all ways best that this should be the fabric of the balloon itself. If made of some form, or at least some proportion of linen, this will probably rend ere long, and, allowing the gas to escape, will soon bring itself to rest. On the other hand, if the balloon proper is a silk one, with sound net and in good condition, it is probable that something else will give way first, and that something may prove to be the hapless passenger or passengers.

And here be it laid down as one first and all-important principle, that in any such awkward predicament as that just described, if there be more than one passenger aboard, let none attempt to get out. In the first place, he may very probably break a limb in so doing, inasmuch as the tangle of the ropes will not allow of his getting cut readily; or, when actually on the ground, he may be caught and impaled by the anchor charging and leaping behind. But, worse than all, he may, in any case, jeopardise the lives of his companions, who stand in need of all the available weight and help that the car contains up to the moment Of coming to final rest.

We have already touched on the early notions as to the means of steering a balloon. Oars had been tested without satisfactory result, and the conception of a rotary screw found favour among theorists at this time, the principle being actually tried with success in working models, which, by mechanical means, could be made to flit about in the still air of the lecture room; but the only feasible method advocated was that already alluded to, which depended on the undesirable action of a trail rope dragging over the ground or through water. The idea was, of course, perfectly practical, and was simply analogous to the method adopted by sailors, who, when floating with the stream but without wind, are desirous of gaining "steerage way." While simply drifting with the flood, they are unable to guide their vessel in any way, and this, in practice, is commonly effected by simply propelling the vessel faster than the stream, in which case the rudder at once becomes available. But the same result is equally well obtained by slowing the vessel, and this is easily accomplished by a cable, with a small anchor or other weight attached, dragging below the vessel. This cable is essentially the same as the guide-rope of the older aeronauts.

It is when we come to consider the impressions and sensations described by sky voyagers of by-gone times that we find them curiously at variance with our own. As an instance, we may state that the earth, as seen from a highflying balloon, used to be almost always described as appearing concave, or like a huge basin, and ingenious attempts were made to prove mathematically that this must be so. The laws of refraction are brought in to prove the fact; or, again, the case is stated thus: Supposing the extreme horizon to be seen when the balloon is little more than a mile high, the range of view on all sides will then be, roughly, some eighty miles. If, then, a line were drawn from the aerial observer to this remote distance, that line would be almost horizontal; so nearly so that he cannot persuade himself that his horizon is otherwise than still on a level with his eye; yet the earth below him lies, as it seems, at the bottom of a huge gulf. Thus the whole visible earth appears as a vast bowl or basin. This is extremely ingenious reasoning, and not to be disregarded; but the fact remains that in the experience of the writer and of many others whom he has consulted, there is no such optical illusion as I have just discussed, and to their vision it is impossible to regard the earth as anything but uniformly flat.

Another impression invariably insisted on by early balloonists is that the earth, on quitting it, appears to drop away into an abyss, leaving the voyagers motionless, and this illusion must, indeed, be probably universal. It is the same illusion as the apparent gliding backwards of objects to a traveller in a railway carriage; only in this latter case the rattling and shaking of the carriage helps the mind to grasp the real fact that the motion belongs to the train itself; whereas it is otherwise with a balloon, whose motion is so perfectly smooth as to be quite imperceptible.

Old ideas, formed upon insufficient observations, even if erroneous, were slow to die. Thus it used to be stated that an upper cloud floor adapted itself to the contour of the land over which it rested, giving what Mr. Monck Mason has called a "phrenological estimate" of the character of the earth below; the clouds, "even when under the influence of rapid motion, seeming to accommodate themselves to all variations of form in the surface of the subjacent soil, rising with its prominences and sinking with its depressions."

Probably few aeronauts of the present time will accept the statement.

It used commonly to be asserted, and is so often to this day, that a feeling as of sea-sickness is experienced in balloon travel, and the notion has undoubtedly arisen from the circumstances attending an ascent in a captive balloon. It were well, now that ballooning bids fair to become popular, to disabuse the public mind of such a wholly false idea. The truth is that a balloon let up with a lengthy rope and held captive will, with a fitful breeze, pitch and sway in a manner which may induce all the unpleasant feelings attending a rough passage at sea. It may do worse, and even be borne to earth with a puff of wind which may come unexpectedly, and considerably unsettle the nerves of any holiday passenger. I could tell of a "captive" that had been behaving itself creditably on a not very settled day suddenly swooping over a roadway and down into public gardens, where it lay incontinently along the ground, and then, before the astonished passengers could attempt to alight, it was seized with another mood, and, mounting once again majestically skyward, submitted to be hauled down with all becoming grace and ease. It is owing to their vagaries and want of manageability that, as will be shown, "captives" are of uncertain use in war. On the other hand, a free balloon is exempt from such disadvantages, and at moderate heights not the smallest feeling of nausea is ever experienced. The only unpleasant sensation, and that not of any gravity, ever complained of, is a peculiar tension in the ears experienced in a rapid ascent, or more often, perhaps, in a descent. The cause, which is trivial and easily removed, should be properly understood, and cannot be given in clearer language than that used by Professor Tyndall: — "Behind the tympanic membrane exists a cavity — the drum of the ear — in part crossed by a series of bones, and in part occupied by air. This cavity communicates with the mouth by means of a duct called the Eustachian tube. This tube is generally closed, the air space behind the tympanic membrane being thus cut off from the external air. If, under these circumstances, the external air becomes denser, it will press the tympanic membrane inwards; if, on the other hand, the air on the other side becomes rarer, while the Eustachian tube becomes closed, the membrane will be pressed outwards. Pain is felt in both cases, and partial deafness is experienced.... By the act of swallowing the Eustachian tube is opened, and thus equilibrium is established between the external and internal pressure."

Founded on physical facts more or less correct in themselves, come a number of tales of olden days, which are at least more marvellous than credible, the following serving as an example. The scientific truth underlying the story is the well-known expedient of placing a shrivelled apple under the receiver of an air pump. As the air becomes rarefied the apple swells, smooths itself out, and presently becomes round and rosy as it was in the summer time. It is recorded that on one occasion a man of mature years made an ascent, accompanied by his son, and, after reaching some height, the youth remarked on how young his father was looking. They still continued to ascend, and the same remark was repeated more than once. And at last, having now reached attenuated regions, the son cried in astonishment, "Why, dad, you ought to be at school!" The cause of this remark was that in the rarefied air all the wrinkles had come out of the old man's face, and his cheeks were as chubby as his son's.

This discussion of old ideas should not be closed without mention of a plausible plea for the balloon made by Wise and others on the score of its value to health. Lofty ascents have proved a strain on even robust constitutions—the heart may begin to suffer, or ills akin to mountain sickness may intervene before a height equal to that of our loftiest mountain is reached. But many have spoken of an exhilaration of spirits not inferior to that of the mountaineer, which is experienced, and without fatigue, in sky voyages reasonably indulged in—of a light-heartedness, a glow of health, a sharpened appetite, and the keen enjoyment of mere existence. Nay, it has been seriously affirmed that "more good may be got by the invalid in an hour or two while two miles up on a fine summer's day than is to be gained in an entire voyage from New York to Madeira by sea."

CHAPTER X. THE COMMENCEMENT OF A NEW ERA.

Resuming the roll of progressive aeronauts in England whose labours were devoted to the practical conquest of the air, and whose methods and mechanical achievements mark the road of advance by which the successes of to-day have been obtained, there stand out prominently two individuals, of whom one has already received mention in these pages.

The period of a single life is seldom sufficient to allow within its span the full development of any new departure in art or science, and it cannot, therefore, be wondered at if Charles Green, though reviving and re-modelling the art of ballooning in our own country, even after an exceptionally long and successful career, left that pursuit to which he had given new birth virtually still in its infancy.

The year following that in which Green conducted the famous Nassau voyage we find him experimenting in the same balloon with his chosen friend and colleague, Edward Spencer, solicitor, of Barnsbury, who, only nine years later, compiles memoranda of thirty-four ascents, made under every variety of circumstance, many being of a highly enterprising nature. We find him writing enthusiastically of the raptures he experienced when sailing over London in night hours, of lofty ascents and extremely low temperatures, of speeding twenty-eight miles in twenty minutes, of grapnel ropes breaking, and of a cross-country race of four miles through woods and hedges. Such was Mr. Spencer the elder, and if further evidence were needed of his practical acquaintance with, as well as personal devotion to, his adopted profession of aeronautics, we have it in the store of working calculations and other minutiae of the craft, most carefully compiled in manuscript by his own hand; these memoranda being to this day constantly consulted by his grandsons, the present eminent aeronauts, Messrs. Spencer Brothers, as supplying a manual of reliable data for the execution of much of the most important parts of their work.

In the terrific ordeal and risk entailed by the daring and fatal parachute descent of Cocking, Green required an assistant of exceptional nerve and reliability, and, as has been recorded, his choice at once fell on Edward Spencer. In this choice it has already been shown that he was well justified, and in the trying circumstances that ensued Green frankly owns that it was his competent companion who was the first to recover himself. A few years later, when a distinguished company, among whom were Albert Smith and Shirley Brooks, made a memorable ascent from Cremorne, Edward Spencer is one of the select party.

Some account of this voyage should be given, and it need not be said that no more graphic account is to be found than that given by the facile pen of Albert Smith himself. His personal narrative also forms an instructive contrast to another which he had occasion to give to the world shortly afterwards, and which shall be duly noticed. The enthusiastic writer first describes, with apparent pride, the company that ascended with him. Besides Mr. Shirley Brooks, there were Messrs. Davidson, of the Garrick Club; Mr. John Lee, well known in theatrical circles; Mr. P. Thompson, of Guy's Hospital, and others—ten in all, including Charles Green as skipper, and Edward Spencer, who, sitting in the rigging, was entrusted with the all-important management of the valve rope.

"The first sensation experienced," Albert Smith continues, "was not that we were rising, but that the balloon remained fixed, whilst all the world below was rapidly falling away; while the cheers with which they greeted our departure grew fainter, and the cheerers themselves began to look like the inmates of many sixpenny Noah's Arks grouped upon a billiard table.... Our hats would have held millions.... And most strange is the roar of the city as it comes surging into the welkin as though the whole metropolis cheered you with one voice.... Yet none beyond the ordinary passengers are to be seen. The noise is as inexplicable as the murmur in the air at hot summer noontide."

The significance of this last remark will be insisted on when the writer has to tell his own experiences aloft over London, as also a note to the effect that there were seen "large enclosed fields and gardens and pleasure grounds where none were supposed to exist by ordinary passengers." Another interesting note, having reference to a once familiar feature on the river, now disappearing, related to the paddle boats of those days, the steamers making a very beautiful effect, "leaving two long wings of foam behind them similar to the train of a table rocket." Highly suggestive, too, of the experiences of railway travellers in the year 1847 is the account of the alighting, which, by the way, was obviously of no very rude nature. "Every time," says the writer, "the grapnel catches in the ground the balloon is pulled up suddenly with a shock that would soon send anybody from his seat, a jerk like that which occurs when fresh carriages are brought up to a railway train." But the concluding paragraph in this rosy narrative affords another and a very notable contrast to the story which that same writer had occasion to put on record before that same year had passed.

"We counsel everybody to go up in a balloon... In spite of the apparent frightful fragility of cane and network nothing can in reality be more secure... The stories of pressure on the ears, intense cold, and the danger of coming down are all fictions.... Indeed, we almost wanted a few perils to give a little excitement to the trip, and have some notion, if possible, of going up the next time at midnight with fireworks in a thunderstorm, throwing away all the ballast, fastening down the valve, and seeing where the wind will send us."

The fireworks, the thunderstorm, and the throwing away of ballast, all came off on the 15th of the following October, when Albert Smith made his second ascent, this time from Vauxhall Gardens, under the guidance of Mr. Gypson, and accompanied by two fellow-passengers. Fireworks, which were to be displayed when aloft, were suspended on a framework forty feet below the car. Lightning was also playing around as they cast off. The description which Albert Smith gives of London by night as seen from an estimated elevation of 4,000 feet, should be compared with other descriptions that will be given in these pages: —

"In the obscurity all traces of houses and enclosures are lost sight of. I can compare it to nothing else than floating over dark blue and boundless sea spangled with hundreds of thousands of stars. These stars were the lamps. We could see them stretching over the river at the bridges, edging its banks, forming squares and long parallel lines of light in the streets and solitary parks. Further and further apart until they were altogether lost in the suburbs. The effect was bewildering."

At 7,000 feet, one of the passengers, sitting in the ring, remarked that the balloon was getting very tense, and the order was given to "ease her" by opening the top valve. The valve line was accordingly pulled, "and immediately afterwards we heard a noise similar to the escape of steam in a locomotive, and the lower part of the balloon collapsed rapidly, and appeared to fly up into the upper portion. At the same instant the balloon began to fall with appalling velocity, the immense mass of loose silk surging and rustling frightfully over our heads.... retreating up away from us more and more into the head of the balloon. The suggestion was made to throw everything over that might lighten the balloon. I had two sandbags in my lap, which were cast away directly.... There were several large bags of ballast, and some bottles of wine, and these were instantly thrown away, but no effect was perceptible. The wind still appeared to be rushing up past us at a fearful rate, and, to add to the horror, we came among the still expiring discharge of the fireworks which floated in the air, so that little bits of exploded cases and touch-paper, still incandescent, attached themselves to the cordage of the balloon and were blown into sparks.... I presume we must have been upwards of a mile from the earth.... How long we were descending I have not the slightest idea, but two minutes must have been the outside.... We now saw the houses, the roofs of which appeared advancing to meet us, and the next instant, as we dashed by their summits, the words, 'Hold hard!' burst simultaneously from all the party.... We were all directly thrown out of the car along the ground, and, incomprehensible as it now appears to me, nobody was seriously hurt."

But "not so incomprehensible, after all," will be the verdict of all who compare the above narrative with the ascents given in a foregoing account of how Wise had fared more than once when his balloon had burst. For, as will be readily guessed, the balloon had in this case also burst, owing to the release of the upper valve being delayed too long, and the balloon had in the natural way transformed itself into a true parachute. Moreover, the fall, which, by Albert Smith's own showing, was that of about a mile in two minutes, was not more excessive than one which will presently be recorded of Mr. Glaisher, who escaped with no material injury beyond a few bruises.

One fact has till now been omitted with regard to the above sensational voyage, namely, the name of the passenger who, sitting in the ring, was the first to point out the imminent danger of the balloon. This individual was none other than Mr. Henry Coxwell, the second, indeed, of the two who were mentioned in the opening paragraph of this chapter as marking the road of progress which it is the scope of these pages to trace, and to whom we must now formally introduce our readers.

This justly famous sky pilot, whose practical acquaintance with ballooning extends over more than forty years, was the son of a naval officer residing near Chatham, and in his autobiography he describes enthusiastically how, a lad of nine years old, he watched through a sea telescope a balloon, piloted by Charles Green, ascend from Rochester and, crossing the Thames, disappear in distance over the Essex flats. He goes on to describe how the incident started him in those early days on boyish endeavours to construct fire balloons and paper parachutes. Some years later his home, on the death of his father, being transferred to Eltham, he came within frequent view of such balloons as, starting from the neighbourhood of London, will through the summer drift with the prevailing winds over that part of Kent. And it was here that, ere long, he came in at the death of another balloon of which Green was in charge.

And from this time onwards the schoolboy with the strange hobby was constantly able to witness the flights and even the inflations of those ships of the air, which, his family associations notwithstanding took precedence of all boyish diversions.

His elder brother, now a naval officer, entirely failed to divert his aspirations into other channels, and it was when the boy had completed sixteen summers that an aeronautic enterprise attracted not only his own, but public attention also. It was the building of a mammoth balloon at Vauxhall under the superintendence of Mr. Green. The launching of this huge craft when completed was regarded as so great an occasion that the young Coxwell, who had by this time obtained a commercial opening abroad, was allowed, at his earnest entreaty, to stay till the event had come off, and fifty years after the hardened sky sailor is found describing with a boyish enthusiasm how thirty-six policemen were needed round that balloon; how enormous weights were attached to the cordage, only to be lifted feet above the ground; while the police were compelled to pass their staves through the meshes to prevent the cords cutting their hands. At this ascent Mr. Hollond was a passenger, and by the middle of the following November all Europe was ringing with the great Nassau venture.

Commercial business did not suit the young Coxwell, and at the age of one-and-twenty we find him trying his hand at the profession of surgeon-dentist, not, however, with any prospect of its keeping him from the longing of his soul, which grew stronger and stronger upon him. It was not till the summer of 1844 that Mr. Hampton, giving an exhibition from the White Conduit Gardens, Pentonville, offered the young man, then twenty-five years old, his first ascent.

In after years Coxwell referred to his first sensations in characteristic language, contrasting them with the experiences of the mountaineer. "In Alpine travels," he says, "the process is so slow, and contact with the crust of the earth so palpable, that the traveller is gradually prepared for each successive phase of view as it presents itself. But in the balloon survey, cities, villages, and vast tracts for observation spring almost magically before the eye, and change in aspect and size so pleasingly that bewilderment first and then unbounded admiration is sure to follow."

The ice was now fairly broken, and, not suffering professional duties to be any hindrance, Coxwell began to make a series of ascents under the leadership of two rival balloonists, Gale and Gypson. One voyage made with the latter he describes as leading to the most perilous descent in the annals of aerostation. This was the occasion, given above, on which Albert Smith was a passenger, and which that talented writer describes in his own fashion. He does not, however, add the fact, worthy of being chronicled, that exactly a week after the appalling adventure Gypson and Coxwell, accompanied by a Captain whose name does not transpire, and loaded with twice the previous weight of fireworks, made a perfectly successful night ascent and descent in the same balloon.

It is very shortly after this that we find Coxwell seduced into undertaking for its owners the actual management of a balloon, the property of Gale, and now to be known as the "Sylph." With this craft he practically began his career as a professional balloonist, and after a few preliminary ascents made in England, was told off to carry on engagements in Belgium.

A long series of ascents was now made on the Continent, and in the troubled state of affairs some stirring scenes were visited, not without some real adventure. One occasion attended with imminent risk occurred at Berlin in 1851. Coxwell relates that a Prussian labourer whom he had dismissed for bad conduct, and who almost too manifestly harboured revenge, nevertheless begged hard for a re-engagement, which, as the man was a handy fellow, Coxwell at length assented to. He took up three passengers beside himself, and at an elevation of some 3,000 feet found it necessary to open the valve, when, on pulling the cord, one of the top shutters broke and remained open, leaving a free aperture of 26 inches by 12 inches, and occasioning such a copious discharge of gas that nothing short of a providential landing could save disaster. But the providential landing came, the party falling into the embrace of a fruit tree in an orchard. It transpired afterwards that the labourer had been seen to tamper with the valve, the connecting lines of which he had partially severed.

Returning to England in 1852 Coxwell, through the accidents inseparable from his profession, found himself virtually in possession of the field.

Green, now advanced in years, was retiring from the public life in which he had won so much fame and honour. Gale was dead, killed in an ascent at Bordeaux. Only one aspirant contested the place of public aeronaut—one Goulston, who had been Gale's patron. Before many months, however, he too met with a balloonist's death, being dashed against some stone walls when ascending near Manchester.

It will not be difficult to form an estimate of how entirely the popularity of the balloon was now reestablished in England, from the mere fact that before the expiration of the year Coxwell had been called upon to make thirty-six voyages. Some of these were from Glasgow, and here a certain coincidence took place which is too curious to be omitted. A descent effected near Milngavie took place in the same field in which Sadler, twenty-nine years before, had also descended, and the same man who caught the rope of Mr. Sadler's balloon performed the same service once again for a fresh visitor from the skies.

The following autumn Coxwell, in fulfilling one out of many engagements, found himself in a dilemma which bore resemblance in a slight degree to a far more serious predicament in which the writer became involved, and which must be told in due place. The preparations for the ascent, which was from the Mile End Road, had been hurried, and after finally getting away at a late hour in the evening, it was found that the valve line had got caught in a fold of the silk, and could not be operated. In consequence, the balloon was, of necessity, left to take its own chance through the night, and, after rising to a considerable height, it slowly lost buoyancy during the chilly hours, and, gradually settling, came to earth near Basingstoke, where the voyager, failing to get help or shelter, made his bed within his own car, lying in an open field, as other aeronauts have had to do in like circumstances.

Coxwell tells of a striking phenomenon seen during that voyage. "A splendid meteor was below the car, and apparently about 600 feet distant. It was blue and yellow, moving rapidly in a N.E. direction, and became extinguished without noise or sparks."

CHAPTER XI. THE BALLOON IN THE SERVICE OF SCIENCE.

At this point we must, for a brief while, drop the history of the famous aeronaut whose early career we have been briefly sketching in the last chapter, and turn our attention to a new feature of English ballooning. We have, at last, to record some genuinely scientific ascents, which our country now, all too tardily, instituted. It was the British Association that took the initiative, and the two men they chose for their purpose were both exceptionally qualified for the task they had in hand. The practical balloonist was none other than the veteran Charles Green, now in his sixty-seventh year, but destined yet to enjoy nearly twenty years more of life. The scientific expert was Mr. John Welsh, well fitted for the projected work by long training at Kew Observatory. The balloon which they used is itself worthy of mention, being the great Nassau Balloon of olden fame.

Welsh was quick to realise more clearly than any former experimentalist that on account of the absence of breeze in a free balloon, as also on account of great solar radiation, the indications of thermometers would, without special precautions, be falsified. He therefore invented a form of aspirating thermometer, the earliest to be met with, and far in advance of any that were subsequently used by other scientists. It consisted of a polished tube, in which thermometers were enclosed, and through which a stream of air was forced by bellows.

The difficulty of obtaining really accurate readings where thermometers are being quickly transported through varying temperatures is generally not duly appreciated. In the case of instruments carried m a balloon it should be remembered that the balloon itself conveys, clinging about it, no inconsiderable quantity of air, brought from other levels, while the temperature of its own mass will be liable to affect any thermometer in close neighbourhood. Moreover, any ordinary form of thermometer is necessarily sluggish in action, as may be readily noticed. If, for example, one be carried from a warm room to a cold passage, or vice versa it will be seen that the column moves very deliberately, and quite a long interval will elapse before it reaches its final position, the cause being that the entire instrument, with any stand or mounting that it may have, will have to adapt itself to the change of temperature before a true record will be obtained. This difficulty applies unavoidably to all thermometers in some degree, and the skill of instrument makers has been taxed to reduce the errors to a minimum. It is necessary, in any case, that a constant stream of surrounding air should play upon the instrument, and though this is most readily effected when instruments are carried aloft by kites, yet even thus it is thought that an interval of some minutes has to elapse before any form of thermometer will faithfully record any definite change of temperature. It is on this account that some allowance must be made for observations which will, in due place, be recorded of scientific explorers; the point to be borne in mind being that, as was mentioned in a former chapter, such observations will have to be regarded as giving readings which are somewhat too high in ascents and too low in descents. Two forms of thermometers at extremely simple construction, yet possessed of great sensibility, will be discussed in later chapters.

The thermometers that Welsh used were undoubtedly far superior to any that were devised before his time and it is much to be regretted that they were allowed to fall into disuse. Perhaps the most important stricture on the observations that will have to be recorded is that the observers were not provided with a base station, on which account the value of results was impaired. It was not realised that it was necessary to make observations on the ground to compare with those that were being made at high altitudes.

Welsh made, in all, four ascents in the summer and autumn of 1852 and in his report he is careful to give the highest praise to his colleague, Green, whose control over his balloon he describes as "so complete that none who accompanied him can be otherwise than relieved from all apprehension, and free to devote attention calmly to the work before him."

The first ascent was made at 3.49 p.m. on August the 17th, under a south wind and with clouds covering some three-quarters of the sky. Welsh's first remark significant, and will be appreciated by anyone who has attempted observational work in a balloon. He states naively that "a short time was lost at first in an attempt to put the instruments into more convenient order, and also from the novelty of the situation." Then he mentions an observation which, in the experience of the writer, is a common one. The lowest clouds, which were about 2,500 feet high and not near the balloon, were passed without being noticed; other clouds were passed at different heights; and, finally, a few star-shaped crystals of snow; but the sun shone almost constantly. Little variation occurred in the direction of travel, which averaged thirty-eight miles an hour, and the descent took place at 5.20 p.m. at Swavesey, near Cambridge.

The second ascent took place at 4.43 p.m. on August 26th, under a gentle east wind and a partially obscured sky. The clouds were again passed without being perceived. This was at the height of 3,000 feet, beyond which was very clear sky of deep blue. The air currents up to the limits of 12,000 feet set from varying directions. The descent occurred near Chesham at 7.45 p.m.

The third ascent, at 2.35 p.m. on October the 21st was made into a sky covered with dense cloud masses lying within 3,000 and 3,700 feet. The sun was then seen shining through cirrus far up. The shadow of the balloon was also seen on the cloud, fringed with a glory, and about this time there was seen "stretching for a considerable length in a serpentine course, over the surface of the cloud, a well-defined belt, having the appearance of a broad road."

Being now at 12,000 feet, Green thought it prudent to reconnoitre his position, and, finding they were near the sea, descended at 4.20 p.m. at Rayleigh, in Essex. Some important notes on the polarisation of the clouds were made.

The fourth and final voyage was made in a fast wind averaging fifty knots from the north-east.

Thin scud was met at 1,900 feet, and an upper stratum at 4,500 feet, beyond which was bright sun. The main shift of wind took place just as the upper surface of the first stratum was reached. In this ascent Welsh reached his greatest elevation, 22,930 feet, when both Green and himself experienced considerable difficulty in respiration and much fatigue. The sea being now perceived rapidly approaching, a hasty descent was made, and many of the instruments were broken.

In summarising his results Welsh states that "the temperature of the air decreases uniformly with height above the earth's surface until at a certain elevation, varying on different days, decrease is arrested, and for the space of 2,000 or 3,000 feet the temperature remains nearly constant, or even increases, the regular diminution being again resumed and generally maintained at a rate slightly less rapid than in the lower part of the atmosphere, and commencing from a higher temperature than would have existed but for the interruption noticed." The analysis of the upper air showed the proportion of oxygen and nitrogen to vary scarcely more than at different spots on the earth.

As it is necessary at this point to take leave of the veteran Green as a practical aeronaut, we may here refer to one or two noteworthy facts and incidents relating to his eventful career. In 1850 M. Poitevin is said to have attracted 140,000 people to Paris to look at an exhibition of himself ascending in a balloon seated on horseback, after which Madame Poitevin ascended from Cremorne Gardens in the same manner, the exhibition being intended as a representation of "Europa on a Bull." This, however, was discountenanced by the authorities and withdrawn. The feats were, in reality, merely the repetitions of one that had been conceived and extremely well carried out by Green many years before—as long ago, in fact, as 1828, when he arranged to make an ascent from the Eagle Tavern, City Road, seated on a pony. To carry out his intention, he discarded the ordinary car, replacing it with a small platform, which was provided with places to receive the pony's feet; while straps attached to the hoop were passed under the animal's body, preventing it from lying down or from making any violent movement. This the creature seemed in no way disposed to attempt, and when all had been successfully carried out and an easy descent effected at Beckenham, the pony was discovered eating a meal of beans with which it had been supplied.

Several interesting observations have been recorded by Green on different occasions, some of which are highly instructive from a practical or scientific point of view. On an ascent from Vaux-hall, in which he was accompanied by his friend Spencer and Mr. Rush, he recorded how, as he constantly and somewhat rapidly rose, the wind changed its direction from N.W. through N. to N.E., while he remained over the metropolis, the balloon all the while rotating on its axis. This continual swinging or revolving of the balloon Green considers an accompaniment of either a rapid ascent or descent, but it may be questioned whether it is not merely a consequence of changing currents, or, sometimes, of an initial spin given inadvertently to the balloon at the moment of its being liberated. The phenomenon of marked change which he describes in the upper currents is highly interesting, and tallies with what the writer has frequently experienced over London proper. Such higher currents may be due to natural environment, and to conditions necessarily prevailing over so vast and varied a city, and they may be able to play an all-important part in the dispersal of London smoke or fog. This point will be touched on later. In this particular voyage Green records that as he was rising at the moment when his barometer reached 19 inches, the thermometer he carried registered 46 degrees, while on coming down, when the barometer again marked 19 inches, the same thermometer recorded only 22 degrees. It will not fail to be recognised that there is doubtless here an example of the errors alluded to above, inseparable from readings taken in ascent and descent.

A calculation made by Green in his earlier years has a certain value. By the time he had accomplished 200 ascents he was at pains to compute that he had travelled across country some 6,000 miles, which had been traversed in 240 hours. From this it would follow that the mean rate of travel in aerial voyages will be about twenty-five miles per hour. Towards the end of his career we find it stated by Lieutenant G. Grover, R.E., that "the Messrs. Green, Father and Son, have made between them some 930 ascents, in none of which have they met with any material accident or failure." This is wonderful testimony, indeed, and we may here add the fact that the father took up his own father, then at the age of eighty-three, in a balloon ascent of 1845, without any serious consequences. But it is time that some account should be given of a particular occasion which at least provided the famous aeronaut with an adventure spiced with no small amount of risk. It was on the 5th of July, 1850, that Green ascended, with Rush as his companion, from Vauxhall, at the somewhat late hour of 7.50 p.m., using, as always, the great Nassau balloon. The rate of rise must have been very considerable, and they presently record an altitude of no less than 20,000 feet, and a temperature of 12 degrees below freezing. They were now above the clouds, where all view of earth was lost, and, not venturing to remain long in this situation, they commenced a rapid descent, and on emerging below found themselves sailing down Sea Reach in the direction of Nore Sands, when they observed a vessel. Their chance of making land was, to say the least, uncertain, and Green, considering that his safety lay in bespeaking the vessel's assistance, opened the valve and brought the car down in the water some two miles north of Sheerness, the hour being 8.45, and only fifty-five minutes since the start. The wind was blowing stiffly, and, catching the hollow of the half-inflated balloon, carried the voyagers rapidly down the river, too fast, indeed, to allow of the vessel's overtaking them. This being soon apparent, Green cast out his anchor, and not without result, for it shortly became entangled in a sunken wreck, and the balloon was promptly "brought up," though struggling and tossing in the broken water. A neighbouring barge at once put off a boat to the rescue, and other boats were despatched by H.M. cutter Fly, under Commander Gurling. Green and Rush were speedily rescued, but the balloon itself was too restive and dangerous an object to approach with safety. At Green's suggestion, therefore, a volley of musketry was fired into the silk' after which it became possible to pass a rope around it and expel the gas. Green subsequently relates how it took a fortnight to restore the damage, consisting of sixty-two bullet rents and nineteen torn gores.

Green's name will always be famous, if only for the fact that it was he who first adopted the use of coal gas in his calling. This, it will be remembered, was in 1821, and it should be borne in mind that at that time household gas had only recently been introduced. In point of fact, it first lighted Pall Mall in 1805, and it was not used for the general lighting of London till 1814.

We are not surprised to find that the great aeronaut at one time turned his attention to the construction of models, and this with no inconsiderable success. A model of his was exhibited in 1840 at the Polytechnic Institution, and is described in the Times as consisting of a miniature balloon of three feet diameter, inflated with coal gas. It was acted on by fans, which were operated by mechanism placed in the car. A series of three experiments was exhibited. First, the balloon being weighted so as to remain poised in the still air of the building, the mechanism was started, and the machine rose steadily to the ceiling. The fans were then reversed, when the model, equally gracefully, descended to the floor. Lastly, the balloon, with a weighted trail rope, being once more balanced in mid-air, the fans were applied laterally, when the machine would take a horizontal flight, pulling the trail rope after it, with an attached weight dragging along the floor until the mechanism had run down, when it again remained stationary. The correspondent of the Times continues, "Mr. Green states that by these simple means a voyage across the Atlantic may be performed in three or four days, as easily as from Vauxhall Gardens to Nassau."

We can hardly attribute this statement seriously to one who knew as well as did Green how fickle are the winds, and how utterly different are the conditions between the still air of a room and those of the open sky. His insight into the difficulties of the problem cannot have been less than that of his successor, Coxwell, who, as the result of his own equally wide experience, states positively, "I could never imagine a motive power of sufficient force to direct and guide a balloon, much less to enable a man or a machine to fly." Even when modern invention had produced a motive power undreamed of in the days we are now considering, Coxwell declares his conviction that inherent difficulties would not be overcome "unless the air should invariably remain in a calm state."

It would be tedious and scarcely instructive to inquire into the various forms of flying machines that were elaborated at this period; but one that was designed in America by Mr. Henson, and with which it was seriously contemplated to attempt to cross the Atlantic, may be briefly described. In theory it was supposed to be capable of being sustained in the air by virtue of the speed mechanically imparted to it, and of the angle at which its advancing under surface would meet the air. The inventor claimed to have produced a steam engine of extreme lightness as well as efficiency, and for the rest his machine consisted of a huge aero-plane propelled by fans with oblique vanes, while a tail somewhat resembling that of a bird was added, as also a rudder, the functions of which were to direct the craft vertically and horizontally respectively. Be it here recorded that the machine did not cross the Atlantic.

One word as to the instruments used up to this time for determining altitudes. These were, in general, ordinary mercurial barometers, protected in various ways. Green encased his instrument in a simple metal tube, which admitted of the column of mercury being easily read. This instrument, which is generally to be seen held in his hand in Green's old portraits, might be mistaken for a mariner's telescope. It is now in the possession of the family of Spencers, the grandchildren of his old aeronautical friend and colleague, and it is stated that with all his care the glass was not infrequently broken in a descent.

Wise, with characteristic ingenuity, devised a rough-and-ready height instrument, which he claims to have answered well. It consisted simply of a common porter bottle, to the neck of which was joined a bladder of the same capacity. The bottle being filled with air of the density of that on the ground, and the bladder tied on in a collapsed state, the expansion of the air in the bottle would gradually fill the bladder as it rose into the rarer regions of the atmosphere. Experience would then be trusted to enable the aeronaut to judge his height from the amount of inflation noticeable in the bladder.

CHAPTER XII. HENRY COXWELL AND HIS CONTEMPORARIES.

Mention should be made in these pages of a night sail of a hundred miles, boldly carried out in 1849 by M. Arban, which took the voyager from Marseilles to Turin fairly over the Alps. The main summit was reached at 11 p.m., when the "snow, cascades, and rivers were all sparkling under the moon, and the ravines and rocks produced masses of darkness which served as shadows to the gigantic picture." Arban was at one time on a level with the highest point of Mont Blanc, the top of which, standing out well above the clouds, resembled "an immense block of crystal sparkling with a thousand fires."

In London, in the year of the Great Exhibition, and while the building was still standing in Hyde Park, there occurred a balloon incident small in itself, but sufficient to cause much sensation at the crowded spot where it took place. The ascent was made from the Hippodrome by Mr. and Mrs. Graham in very boisterous weather, and, on being liberated, the balloon seems to have fouled a mast, suffering a considerable rent. After this the aeronauts succeeded in clearing the trees in Kensington Gardens, and in descending fairly in the Park, but, still at the mercy of the winds, they were carried on to the roof of a house in Arlington Street, and thence on to another in Park Place, where, becoming lodged against a stack of chimneys, they were eventually rescued by the police without any material damage having been done.

But this same summer saw the return to England of Henry Coxwell, and for some years the story of the conquest of the air is best told by following his stirring career, and his own comments on aeronautical events of this date. We find him shortly setting about carrying out some reconnoitring and signalling experiments, designed to be of use in time of war. This was an old idea of his, and one which had, of course, been long entertained by others, having, indeed, been put to some practical test in time of warfare. It will be well to make note of what attention the matter had already received, and of what progress had been made both in theory and practice.

We have already made some mention in Chapter IV. of the use which the French had made of balloons in their military operations at the end of the eighteenth and beginning of nineteenth the century. It was, indeed, within the first ten years after the first invention of the balloon that, under the superintendence of the savants of the French Academy, a practical school of aeronautics was established at Meudon. The names of Guyton, De

Morveau (a distinguished French chemist), and Colonel Coutelle are chiefly associated with the movement, and under them some fifty students received necessary training. The practising balloon had a capacity of 17,000 cubic feet, and was inflated with pure hydrogen, made by what was then a new process as applied to ballooning, and which will be described in a future chapter. It appears that the balloon was kept always full, so that any opportunity of calm weather would be taken advantage of for practice. And it is further stated that a balloon was constructed so sound and impervious that after the lapse of two months it was still capable, without being replenished, of raising into the air two men, with necessary ballast and equipment. The practical trial for the balloon in real service came off in June, 1794, when Coutelle in person, accompanied by two staff officers, in one of the four balloons which the French Army had provided, made an ascent to reconnoitre the Austrian forces at Fleurus. They ascended twice in one day, remaining aloft for some four hours, and, on their second ascent being sighted, drew a brisk fire from the enemy. They were unharmed, however, and the successful termination of the battle of Fleurus has been claimed as due in large measure to the service rendered by that balloon.

The extraordinary fact that the use of the balloon was for many years discontinued in the French Army is attributed to a strangely superstitious prejudice entertained by Napoleon. Las Cases (in his "Private Life of Napoleon at St. Helena ") relates an almost miraculous story of Napoleon's coronation. It appears that a sum of 23,500 francs was given to M. Garnerin to provide a balloon ascent to aid in the celebrations, and, in consequence, a colossal machine was made to ascend at 11 p.m. on December 16th from the front of Notre Dame, carrying 3,000 lights. This balloon was unmanned, and at its departure apparently behaved extremely well, causing universal delight. During the hours of darkness, however, it seems to have acquitted itself in a strange and well-nigh preternatural manner, for at daybreak it is sighted on the horizon by the inhabitants of Rome, and seen to be coming towards their city. So true was its course that, as though with predetermined purpose, it sails on till it is positively over St. Peter's and the Vatican, when, its mission being apparently fulfilled, it settles to earth, and finally ends its career in the Lake Bracciano. Regarded from whatever point of view, the flight was certainly extraordinary, and it is not surprising that in that age it was regarded as nothing less than a portent. Moreover, little details of the wonderful story were quickly endowed with grave significance. The balloon on reaching the ground rent itself. Next, ere it plunged into the water, it carefully deposited a portion of its crown on the tomb of Nero. Napoleon, on learning the facts, forbade that they should ever be referred to. Further, he thenceforward discountenanced the balloon in his army, and the establishment at Meudon was abandoned.

There is record of an attempt of some sort that was made to revive the French military ballooning school in the African campaign of 1830, but it was barren of results. Again, it has been stated that the Austrians used balloons for reconnaissance, before Venice in 1849, and yet again the same thing is related of the Russians at the time of the siege of Sebastopol, though Kinglake does not mention the circumstance. In 1846 Wise drew up and laid before the American War Office an elaborate scheme for the reduction of Vera Cruz. This will be discussed in its due place, though it will be doubtless considered as chimerical.

On the other hand, eminently practical were the experiments co-ordinated and begun to be put to an actual test by Mr. Coxwell, who, before he could duly impress his project upon the military authorities, had to make preliminary trials in private ventures. The earliest of these was at the

Surrey Zoological Gardens in the autumn of 1854, and it will be granted that much ingenuity and originality were displayed when it is considered that at that date neither wireless telegraphy, electric flashlight, nor even Morse Code signalling was in vogue. According to his announcement, the spectators were to regard his balloon, captive or free, as floating at a certain altitude over a beleaguered fortress, the authorities in communication with it having the key of the signals and seeking to obtain through these means information as to the approach of an enemy. It was to be supposed that, by the aid of glasses, a vast distance around could be subjected to careful scrutiny, and a constant communication kept up with the authorities in the fortress. Further, the flags or other signals were supposed preconcerted and unknown to the enemy, being formed by variations of shape and colour. Pigeons were also despatched from a considerable height to test their efficiency under novel conditions. The public press commented favourably on the performance and result of this initial experiment.

Mr. Coxwell's account of an occasion when he had to try conclusions with a very boisterous wind, and of the way in which he negotiated a very trying and dangerous landing, will be found alike interesting and instructive. It was an ascent from the Crystal Palace, and the morning was fair and of bright promise outwardly; but Coxwell confesses to have disregarded a falling glass. The inflation having been progressing satisfactorily, he retired to partake of luncheon, entirely free from apprehensions; but while thus occupied, he was presently sought out and summoned by a gardener, who told him that his balloon had torn away, and was now completely out of control, dragging his men about the bushes. On reaching the scene, the men, in great strength, were about to attempt a more strenuous effort to drag the balloon back against the wind, which Coxwell promptly forbade, warning them that so they would tear all to pieces. He then commenced, as it were, to "take in a reef," by gathering in the slack of the silk, which chiefly was catching the wind, and by drawing in the net, mesh by mesh, until the more inflated portion of the balloon was left snug and offering but little resistance to the gale, when he got her dragged in a direction slanting to the wind and under the lee of trees.

Eventually a hazardous and difficult departure was effected, Mr. Chandler, a passenger already booked, insisting on accompanying the aeronaut, in spite of the latter's strongest protestations. And their first peril came quickly, in a near shave of fouling the balcony of the North Tower, which they avoided only by a prompt discharge of sand, the crowd cheering loudly as they saw how the crisis was avoided. The car, adds Mr. Coxwell in his memoirs, "was apparently trailing behind the balloon with a pendulous swing, which is not often the case... In less than two minutes we entered the lower clouds, passing through them quickly, and noticing that their tops, which are usually of white, rounded conformation, were torn into shreds and crests of vapour. Above, there was a second wild-looking stratum of another order. We could hear, as we hastened on, the hum of the West End of London; but we were bowling along, having little time to look about us, though some extra sandbags were turned to good account by making a bed of them at the bottom ends of the car, which we occupied in anticipation of a rough landing."

As it came on to rain hard the voyagers agreed to descend, and Coxwell, choosing open ground, succeeded in the oft-attempted endeavour to drop his grapnel in front of a bank or hedge-row. The balloon pulled up with such a shock as inevitably follows when flying at sixty miles an hour, and Mr. Coxwell continues:—"We were at this time suspended like a kite, and it was not so much the quantity of gas which kept us up as the hollow surface of loose silk, which acted like a falling kite, and the obvious game of skill consisted in not letting out too much gas to make the balloon pitch heavily with a thud that would have been awfully unpleasant; but to jockey our final touch in a gradual manner, and yet to do it as quickly as possible for fear of the machine getting adrift, since, under the peculiar circumstances in which we were placed, it would have inevitably fallen with a crushing blow, which might have proved fatal. I never remember to have been in a situation when more coolness and nicety were required to overcome the peril which here beset us; while on that day the strong wind was, strange as it may sound, helping us to alight easily, that is to say as long as the grapnel held fast and the balloon did not turn over like an unsteady kite." Such peril as there was soon terminated without injury to either voyager.

The same remark will apply to an occasion when Coxwell was caught in a thunderstorm, which he thus describes in brief: — "On a second ascent from Chesterfield we were carried into the midst of gathering clouds, which began to flash vividly, and in the end culminated in a storm. There were indications, before we left the earth, as to what might be expected. The lower breeze took us in another direction as we rose, but a gentle, whirling current higher up got us into the vortex of a highly charged cloud.... We had to prove by absolute experience whether the balloon was insulated and a non-conductor. Beyond a drenching, no untoward incident occurred during a voyage lasting in all three-quarters of an hour."

A voyage which Coxwell (referring, doubtless, to aerial travel over English soil only) describes as "being so very much in excess of accustomary trips in balloons" will be seen to fall short of one memorable voyage of which the writer will have to give his own experiences. Some account, however, of what the famous aeronaut has to tell will find a fitting place here.

It was an ascent on a summer night from North Woolwich, and on this occasion Coxwell was accompanied by two friends, one being Henry Youens, who subsequently became a professional balloonist of considerable repute, and who at this time was an ardent amateur. It was half an hour before midnight when the party took their places, and, getting smartly away from the crowd in the gala grounds, shot over the river, and shortly were over the town of Greenwich with the lights of London well ahead. Then their course took them over Kennington Oval, Vauxhall Bridge, and Battersea, when they presently heard the strains of a Scotch polka. This came up from the then famous Gardens of Cremorne, and, the breeze freshening, it was but a few minutes later when they stood over Kingston, by which time it became a question whether, being now clear of London, they should descend or else live out the night and take what thus might come their way. This course, as the most prudent, as well as the most fascinating, was that which commended itself, and at that moment the hour of midnight was heard striking, showing that a fairly long distance had been covered in a short interval of time.

From this period they would seem to have lost their way, and though scattered lights were sighted ahead, they were soon in doubt as to whether they might not already be nearing the sea, a doubt that was strengthened by their hearing the cry of sea-fowl. After a pause, lights were seen looming under the haze to sea-ward, which at times resembled water; and a tail like that of a comet was discerned, beyond which was a black patch of considerable size.

The patch was the Isle of Wight, and the tail the Water from Southampton. They were thus wearing more south and towards danger. They had no Davy lamp with which to read their aneroid, and could only tell from the upward flight of fragments of paper that they were descending. Another deficiency in their equipment was the lack of a trail rope to break their fall, and for some time they were under unpleasant apprehension of an unexpected and rude impact with the ground, or collision with some undesirable object. This induced them to discharge sand and to risk the consequences of another rise into space, and as they mounted they were not reassured by sighting to the south a ridge of lighter colour, which strongly suggested the coast line.

But it was midsummer, and it was not long before bird life awakening was heard below, and then a streak of dawn revealed their locality, which was over the Exe, with Sidmouth and Tor Bay hard by on their left. Then from here, the land jutting seawards, they confidently traversed Dartmoor, and effected a safe, if somewhat unseasonable, descent near Tavistock. The distance travelled was considerable, but the duration, on the aeronaut's own showing, was less than five hours.

In the year 1859 the Times commented on the usefulness of military balloons in language that fully justified all that Coxwell had previously claimed for them. A war correspondent, who had accompanied the Austrian Army during that year, asks pertinently how it had happened that the French had been ready at six o'clock to make a combined attack against the Austrians, who, on their part, had but just taken up positions on the previous evening. The correspondent goes on to supply the answer thus:—"No sooner was the first Austrian battalion out of Vallegio than a balloon was observed to rise in the air from the vicinity of Monsambano—a signal, no doubt, for the French in Castiglione. I have a full conviction that the Emperor of the French knew overnight the exact position of every Austrian corps, while the Emperor of Austria was unable to ascertain the number or distribution of the forces of the allies."

It appears that M. Godard was the aeronaut employed to observe the enemy, and that fresh balloons for the French Army were proceeded with.

The date was now near at hand when Coxwell, in partnership with Mr. Glaisher, was to take part in the classical work which has rendered their names famous throughout the world. Before proceeding to tell of that period, however, Mr. Coxwell has done well to record one aerial adventure, which, while but narrowly missing the most serious consequences, gives a very practical illustration of the chances in favour of the aeronaut under extreme circumstances.

It was an ascent at Congleton in a gale of wind,
a and the company of two passengers—Messrs.
Pearson, of Lawton Hall—was pressed upon him.
Everything foretold a rough landing, and some
time after the start was made the outlook was not
improved by the fact that the dreaded county of
Derbyshire was seen approaching; and it was
presently apparent that the spot on which they
had decided to descend was faced by rocks and a
formidable gorge. On this, Coxwell attempted to
drop his grapnel in front of a stone wall, and so
far with success; but the wall went down, as also
another and another, the wicker car passing, with
its great impetus, clean through the solid obstacles, till at last the balloon slit from top to bottom.
Very serious injuries to heads and limbs were
sustained, but no lives were lost, and Coxwell
himself, after being laid up at Buxton, got home
on crutches.

CHAPTER XIII. SOME NOTEWORTHY ASCENTS.

It was the year 1862, and the scientific world in England determined once again on attempting observational work in connection with balloons. There had been a meeting of the British Association at Wolverhampton, and, under their auspices, and with the professional services of Thomas Lythgoe, Mr. Creswick, of Greenwich Observatory, was commissioned to make a lofty scientific ascent with a Cremorne balloon. The attempt, however, was unsatisfactory; and the balloon being condemned, an application was made to Mr. Coxwell to provide a suitable craft, and to undertake its management. The principals of the working committee were Colonel Sykes, M.P., Dr. Lee, and Mr. James Glaisher, F.R.S., and a short conference between these gentlemen and the experienced aeronaut soon made it clear that a mammoth balloon far larger than any in existence was needed for the work in hand. But here a fatal obstacle presented itself in lack of funds, for it transpired that the grant voted was only to be devoted to trial ascents.

It was then that Mr. Coxwell, with characteristic enterprise, undertook, at his own cost, to build a suitable balloon, and, moreover, to have it ready by Midsummer Day. It was a bold, as well as a generous, offer; for it was now March, and, according to Mr. Coxwell's statement, if silk were employed, the preparation and manufacture would occupy six months and cost not less than L2,000. The fabric chosen was a sort of American cloth, and by unremitting efforts the task was performed to time, and the balloon forwarded to Wolverhampton, its dimensions being 55 feet in diameter, 80 feet in height from the ground, with a capacity of 93,000 cubic feet. But the best feature in connection with it was the fact that Mr. Glaisher himself was to make the ascents as scientific observer.

No time was lost in getting to work, but twice over the chosen days were unsuitable, and it was not till July 17th that the two colleagues, of whom so much is to be told, got away at 9.30 a.m. with their balloon only two-thirds full, to allow of expansion to take place in such a lofty ascent as was contemplated. And, when it is considered that an altitude of five miles was reached, it will be granted that the scientific gentleman who was making his maiden ascent that day showed remarkable endurance and tenacity of purpose—the all-important essential for the onerous and trying work before him. At 9.56 the balloon had disappeared from sight, climbing far into the sky in the E.N.E. The story of the voyage we must leave in Mr. Glaisher's hands. Certain events, however, associated with other aeronauts, which had already happened, and which should be considered in connection with the new drama now to be introduced, may fittingly here meet with brief mention.

The trouble arising from the coasting across country of a fallen and still half-inflated balloon has already been sufficiently illustrated, and needs little further discussion. It is common enough to see a balloon, when full and round, struggling restively under a moderate breeze with a score of men, and dragging them, and near a ton of sand-bags as well, about the starting ground. But, as has already been pointed out, the power of the wind on the globe is vastly increased when the silk becomes slack and forms a hollow to hold the wind, like a bellying sail. Various means to deal with this difficulty have been devised, one of these being an emergency, or ripping valve, in addition to the ordinary valve, consisting of an arrangement for tearing a large opening in the upper part of one of the gores, so that on reaching earth the balloon may be immediately crippled and emptied of so large a quantity of gas as to render dragging impossible. Such a method is not altogether without drawbacks, one of these being the confusion liable to arise from there being more than one valve line to reckon with. To obviate this, it has been suggested that the emergency line should be of a distinctive colour.

But an experiment with a safeguard to somewhat of this nature was attended with fatal consequence in the year 1824. A Mr. Harris, a lieutenant in the British Navy, ascended from the Eagle Tavern, City Road, with a balloon fitted with a contrivance of his own invention, consisting of a large hinged upper valve, having within it a smaller valve of the same description, the idea being that, should the operation of the smaller outlet not suffice for any occasion, then the shutter of the larger opening might be resorted to, to effect a more liberal discharge of gas.

Mr. Harris took with him a young lady, Miss Stocks by name, and apparently the afternoon—it being late May—was favourable for an aerial voyage; for, with full reliance on his apparatus, he left his grapnel behind, and was content with such assistance as the girl might be able to render him. It was not long before the balloon was found descending, and with a rapidity that seemed somewhat to disturb the aeronaut; and when, after a re-ascent, effected by a discharge of ballast, another decided downward tendency ensued, Mr. Harris clearly realised that something was wrong, without, however, divining the cause. The story subsequently told by the girl was to the effect that when the balloon was descending the second time she was spoken to by her unfortunate companion in an anxious manner. "I then heard the balloon go 'Clap! clap!' and Mr. Harris said he was afraid it was bursting, at which I fainted, and knew no more until I found myself in bed." A gamekeeper tells the sequel, relating that he observed the balloon, which was descending with great velocity, strike and break the head of an oak tree, after which it also struck the ground. Hurrying up, he found the girl insensible, and Mr. Harris already dead, with his breast bone and several ribs broken. The explanation of the accident given by Mr. Edward Spencer is alike convincing and instructive. This eminently practical authority points out that the valve lines must have been made taut to the hoop at the time that the balloon was full and globular. Thus, subsequently, when from diminution of gas the balloon's shape elongated, the valve line would become strained and begin to open the valve, but in such a gradual manner as to escape the notice of the aeronaut. Miss Stocks, far from being unnerved by the terrible experience, actually made three subsequent ascents in company with Mr. Green.

It deserves mention that another disaster, equally instructive, but happily not attended with loss of life, occurred in Dublin in 1844 to Mr. Hampton, who about this time made several public and enterprising voyages. He evidently was possessed of admirable nerve and decision, and did not hesitate to make an ascent from the Porto-Bello Gardens in face of strong wind blowing seawards, and in spite of many protestations from the onlookers that he was placing himself in danger. This danger he fully realised, more particularly when he recognised that the headland on which he hoped to alight was not in the direction of the wind's course. Resolved, however, on gratifying the crowd, Mr. Hampton ascended rapidly, and then with equal expedition commenced a precipitate descent, which he accomplished with skill and without mishap. But the wind was still boisterous, and the balloon sped onward along the ground towards fresh danger unforeseen, and perhaps not duly reckoned with. Ahead was a cottage, the chimney of which was on fire. A balloonist in these circumstances is apt to think little of a single small object in his way, knowing how many are the chances of missing or of successfully negotiating any such obstacle. The writer on one occasion was, in the judgment of onlookers below, drifting in dangerous proximity to the awful Cwmavon stack in Glamorganshire, then in full blast; yet it was a fact that that vast vent of flame and smoke passed almost unheeded by the party in the descending car. It may have been thus, also, with Mr. Hampton, who only fully realised his danger when his balloon blew up "with an awfully grand explosion," and he was reduced to the extremity of jumping for his life, happily escaping the mass of burning silk and ropes.

The awful predicament of falling into the sea, which has been illustrated already, and which will recur again in these pages, was ably and successfully met by Mr. Cunningham, who made an afternoon ascent from the Artillery Barracks at Clevedon, reaching Snake Island at nightfall, where, owing to the gathering darkness, he felt constrained to open his valve. He quickly commenced descending into the sea, and when within ten feet of the water, turned the "detaching screw" which connected the car with the balloon. The effect of this was at once to launch him on the waves, but, being still able to keep control over the valve, he allowed just enough gas to remain within the silk to hold the balloon above water. He then betook himself to the paddles with which his craft was provided, and reached Snake Island with the balloon in tow. Here he seems to have found good use for a further portion of his very complete equipment; for, lighting a signal rocket, he presently brought a four-oared gig to his succour from Portsmouth Harbour.

The teaching of the above incident is manifest enough. If it should be contemplated to use the balloon for serious or lengthened travel any-where within possible reach of the sea-board — and this must apply to all parts of the British Isles — it must become a wise precaution, if not an absolute necessity, to adopt some form of car that would be of avail in the event of a fall taking place in the sea. Sufficient confirmation of this statement will be shortly afforded by a memorable voyage accomplished during the partnership of Messrs. Glaisher and Coxwell, one which would certainly have found the travellers in far less jeopardy had their car been convertible into a boat. We have already seen how essential Wise considered this expedient in his own bolder schemes, and it may further be mentioned here that modern air ships have been designed with the intention of making the water a perfectly safe landing.

The ballooning exploits which, however, we have now to recount had quite another and more special object consistently in view — that of scientific investigation; and we would here premise that the proper appreciation of these investigations will depend on a due understanding of the attendant circumstances, as also of the constant characteristic behaviour of balloons, whether despatched for mere travel or research.

First let us regard the actual path of a balloon in space when being manoeuvred in the way we read of in Mr. Glaisher's own accounts. This part is in most cases approximately indicated in that most attractive volume of his entitled, "Travels in the Air," by diagrams giving a sectional presentment of his more important voyages; but a little commonplace consideration may take the place of diagrams.

It has been common to assert that a balloon poised in space is the most delicate balance conceivable. Its intrinsic weight must be exactly equal to the weight of the air it displaces, and since the density of the air decreases according to a fixed law, amounting, approximately, to a difference in barometric reading of 0.1 inch for every 90 feet, it follows, theoretically, that if a balloon is poised at 1,000 feet above sea level, then it would not be in equilibrium at any other height, so long as its weight and volume remain the same. If it were 50 feet higher it must commence descending, and, if lower, then it must ascend till it reaches its true level; and, more than that, in the event of either such excursion mere impetus would carry it beyond this level, about which it would oscillate for a short time, after the manner of the pendulum. This is substantially true, but it must be taken in connection with other facts which have a far greater influence on a balloon's position or motion.

For instance, in the volume just referred to it is stated by M. Gaston Tissandier that on one occasion when aloft he threw overboard a chicken bone, and, immediately consulting a barometer, had to admit on "clearest evidence that the bone had caused a rise of from twenty to thirty yards, so delicately is a balloon equipoised in the air." Here, without pausing to calculate whether the discharge of an ounce or so would suffice to cause a large balloon to ascend through ninety feet, it may be pointed out that the record cannot be trustworthy, from the mere fact that a free balloon is from moment to moment being subjected to other potent influences, which necessarily affect its position in space. In daytime the sun's influence is an all-important factor, and whether shining brightly or partially hidden by clouds, a slight difference in obscuration will have a ready and marked effect on the balloon's altitude. Again, a balloon in transit may pass almost momentarily from a warmer layer of air to a colder, or vice versa, the plane of demarcation between the two being very definite and abrupt, and in this case altitude is at once affected; or, yet again, there are the descending and ascending currents, met with constantly and unexpectedly, which have to be reckoned with.

Thus it becomes a fact that a balloon's vertical course is subjected to constant checks and vicissitudes from a variety of causes, and these will have to be duly borne in mind when we are confronted with the often surprising results and readings which are supplied by scientific observers. With regard to the close proximity, without appreciable intermingling, of widely differing currents, it should be mentioned that explorers have found in regions where winds of different directions pass each other that one air stream appears actually to drag against the surface of the other, as though admitting no interspace where the streams might mingle. Indeed, trustworthy observers have stated that even a hurricane can rage over a tranquil atmosphere with a sharply defined surface of demarcation between calm and storm. Thus, to quote the actual words of Charles Darwin, than whom it is impossible to adduce a more careful witness, we find him recording how on mountain heights he met with winds turbulent and unconfined, yet holding courses "like rivers within their beds."

It is in tracing the trend of upper air streams, to whose wayward courses and ever varying conditions we are now to be introduced, that much of our most valuable information has come, affecting the possibility of forecasting British wind and weather. It should need no insisting on that the data required by meteorologists are not sufficiently supplied by the readings of instruments placed on or near the ground, or by the set of the wind as determined by a vane planted on the top of a pole or roof of a building. The chief factors in our meteorology are rather those broader and deeper conditions which obtain in higher regions necessarily beyond our ken, until those regions are duly and diligently explored.

Mr. Glaisher's estimate of the utility of the balloon as an instrument of research, formed at the conclusion of his aeronautical labours, has a special value and significance. Speaking with all the weight attaching to so trained and eminent an observer, he declares, "The balloon, considered as an instrument for vertical exploration, presents itself to us under a variety of aspects, each of which is fertile in suggestions. Regarding the atmosphere as the great laboratory of changes which contain the germ of future dis discoveries, to belong respectively, as they unfold, to the chemist and meteorologist, the physical relation to animal life of different heights, the form of death which at certain elevations waits to accomplish its destruction, the effect of diminished pressure upon individuals similarly placed, the comparison of mountain ascents with the experiences of aeronauts, are some of the questions which suggest themselves and faintly indicate enquiries which naturally ally themselves to the course of balloon experiments. Sufficiently varied and important, they will be seen to rank the balloon as a valuable aid to the uses of philosophy, and rescue it from the impending degradation of continuing a toy fit only to be exhibited or to administer to the pleasures of the curious and lovers of adventure."

The words of the same authority as to the possible practical development of the balloon as an aerial machine should likewise be quoted, and will appear almost prophetic. "In England the subject of aero-station has made but little progress, and no valuable invention has arisen to facilitate travelling in the air. In all my ascents I used the balloon as I found it. The desire which influenced me was to ascend to the higher regions and travel by its means in furtherance of a better knowledge of atmospheric phenomena. Neither its management nor its improvement formed a part of my plan. I soon found that balloon travelling was at the mercy of the wind, and I saw no probability of any method of steering balloons being obtained. It even appeared to me that the balloon itself, admirable for vertical ascents, was not necessarily a first step in aerial navigation, and might possibly have no share in the solution of the problem. It was this conviction that led to the formation of the Aeronautical Society a few years since under the presidency of the Duke of Argyll. In the number of communications made to this society it is evident that many minds are taxing their ingenuity to discover a mode of navigating the air; all kinds of imaginary projects have been suggested, some showing great mechanical ingenuity, but all indicating the want of more knowledge of the atmosphere itself. The first great aim of this society is the connecting the velocity of the air with its pressure on plane surfaces at various inclinations.

"There seems no prospect of obtaining this relation otherwise than by a careful series of experiments."

CHAPTER XIV. THE HIGHEST ASCENT ON RECORD.

Mr. Glaisher's instrumental outfit was on an elaborate and costly scale, and the programme of experimental work drawn up for him by the Committee of the British Association did not err on the side of too much modesty. In the first place the temperature and moisture of the atmosphere were to be examined. Observations on mountain sides had determined that thermometers showed a decrease of 1 degree F. for every 300 feet, and the accuracy of this law was particularly to be tested. Also, investigations were to be made as to the distribution of vapour below the clouds, in them, and above them. Then careful observations respecting the dew point were to be undertaken at all accessible heights, and, more particularly, up to those heights where man may be resident or troops may be located. The comparatively new instrument, the aneroid barometer, extremely valuable, if only trustworthy, by reason of its sensibility, portability and safety, was to be tested and compared with the behaviour of a reliable mercurial barometer. Electrical conditions were to be examined; the presence of ozone tested; the vibration of a magnet was again to be resorted to to determine how far the magnetism of the earth might be affected by height. The solar spectrum was to be observed; air was to be collected at different heights for analysis; clouds, also upper currents, were to be reported on. Further observations were to be made on sound, on solar radiation, on the actinic action of the sun, and on atmospheric phenomena in general.

All this must be regarded as a large order where only a very limited number of ascents were contemplated, and it may be mentioned that some of the methods of investigation, as, for instance, the use of ozone papers, would now be generally considered obsolete; while the mechanical aspiration of thermometers by a stream of air, which, as we have pointed out, was introduced by Welsh, and which is strongly insisted on at the present day, was considered unnecessary by Mr. Glaisher in the case of wet and dry bulb hygrometers. The entire list of instruments, as minutely described by the talented observer, numbered twenty-two articles, among which were such irreproachable items as a bottle of water and a pair of scissors.

The following is a condensed account, gathered from Mr. Glaisher's own narrative, of his first ascent, which has been already briefly sketched in these pages by the hand of Mr. Coxwell. Very great difficulties were experienced in the inflation, which operation appeared as if it would never be completed, for a terrible W.S.W. wind was constantly blowing, and the movements of the balloon were so great and so rapid that it was impossible to fix a single instrument in its position before quitting the earth, a position of affairs which, says Mr. Glaisher, "was by no means cheering to a novice who had never before put his foot in the car of a balloon," and when, at last, at 9.42 a.m., Mr. Coxwell cast off, there was no upward motion, the car simply dragging on its side till the expiration of a whole minute, when the balloon lifted, and in six minutes reached the first cloud at an altitude of 4,467 feet. This cloud was passed at 5,802 feet, and further cloud encountered at 2,000 feet further aloft. Four minutes later, the ascent proceeding, the sun shone out brightly, expanding the balloon into a perfect globe and displaying a magnificent view, which, however, the incipient voyager did not allow himself to enjoy until the instruments were arranged in due order, by which time a height of 10,000 feet was recorded.

Mr. Glaisher apparently now had opportunity for observing the clouds, which he describes as very beautiful, and he records the hearing of a band of music at a height of 12,709 feet, which was attained in exactly twenty minutes from the start. A minute later the earth was sighted through a break in the clouds, and at 16,914 feet the clouds were far below, the sky above being perfectly cloudless, and of an intense Prussian blue.

By this time Mr. Glaisher had received his first surprise, as imparted by the record of his instruments. At starting, the temperature of the air had stood at 59 degrees. Then at 4,000 feet this was reduced to 45 degrees; and, further, to 26 degrees at 10,000 feet, when it remained stationary through an ascent of 3,000 feet more, during which period both travellers added to their clothing, anticipating much accession of cold. However, at 15,500 feet the temperature had actually risen to 31 degrees, increasing to no less than 42 degrees at 19,500 feet.

Astonishing as this discovery was, it was not the end of the wonder, for two minutes later, on somewhat descending, the temperature commenced decreasing so rapidly as to show a fall of 27 degrees in 26 minutes. As to personal experiences, Mr. Glaisher should be left to tell his own story. "At the height of 18,844 feet 18 vibrations of a horizontal magnet occupied 26.8 seconds, and at the same height my pulse beat at the rate of 100 pulsations per minute. At 19,415 feet palpitation of the heart became perceptible, the beating of the chronometer seemed very loud, and my breathing became affected. At 19,435 feet my pulse had accelerated, and it was with increasing difficulty that I could read the instruments; the palpitation of the heart was very perceptible; the hands and lips assumed a dark bluish colour, but not the face. At 20,238 feet 28 vibrations of a horizontal magnet occupied 43 seconds. At 21,792 feet I experienced a feeling analogous to sea-sickness, though there was neither pitching nor rolling in the balloon, and through this illness I was unable to watch the instrument long enough to lower the temperature to get a deposit of dew. The sky at this elevation was of a very deep blue colour, and the clouds were far below us. At 22,357 feet I endeavoured to make the magnet vibrate, but could not; it moved through arcs of about 20 degrees, and then settled suddenly.

"Our descent began a little after 11 a.m., Mr. Coxwell experiencing considerable uneasiness at our too close vicinity to the Wash. We came down quickly from a height of 16,300 feet to one of 12,400 feet in one minute; at this elevation we entered into a dense cloud which proved to be no less than 8,000 feet in thickness and whilst passing through this the balloon was invisible from the car. From the rapidity of the descent the balloon assumed the shape of a parachute, and though Mr. Coxwell had reserved a large amount of ballast, which he discharged as quickly as possible, we collected so much weight by the condensation of the immense amount of vapour through which we passed that, notwithstanding all his exertions, we came to the earth with a very considerable shock, which broke nearly all the instruments.... The descent took place at Langham, near Oakham."

Just a month later Mr. Glaisher, bent on a yet loftier climb, made his second ascent, again under Mr. Coxwell's guidance, and again from Wolverhampton. Besides attending to his instruments he found leisure to make other chance notes by the way. He was particularly struck by the beauty of masses of cloud, which, by the time 12,000 feet were reached, were far below, "presenting at times mountain scenes of endless variety and grandeur, while fine dome-like clouds dazzled and charmed the eye with alternations and brilliant effects of light and shade."

When a height of about 20,000 feet had been reached thunder was heard twice over, coming from below, though no clouds could be seen. A height of 4,000 feet more was attained, and shortly after this Mr. Glaisher speaks of feeling unwell. It was difficult to obtain a deposit of dew on the hygrometer, and the working of the aspirator became troublesome. While in this region a sound like that of loud thunder came from the sky. Observations were practically completed at this point, and a speedy and safe return to earth was effected, the landing being at Solihull, seven miles from Birmingham.

It was on the 5th of September following that the same two colleagues carried out an exploit which will always stand alone in the history of aeronautics, namely, that of ascending to an altitude which, based on the best estimate they were able to make, they calculated to be no less than seven miles. Whatever error may have unavoidably come into the actual estimate, which is to some extent conjectural, is in reality a small matter, not the least affecting the fact that the feat in itself will probably remain without a parallel of its kind. In these days, when aeronauts attempt to reach an exceptionally lofty altitude, they invariably provide themselves with a cylinder of oxygen gas to meet the special emergencies of the situation, so that when regions of such attenuated air are reached that the action of heart and lungs becomes seriously affected, it is still within their power to inhale the life-giving gas which affords the greatest available restorative to their energies. Forty years ago, however, cylinders of compressed oxygen gas were not available, and on this account alone we may state without hesitation that the enterprise which follows stands unparalleled at the present hour.

The filling station at Wolverhampton was quitted at 1.3 p.m., the temperature of the air being 59 degrees on the ground, and falling to 41 degrees at an altitude of 5,000 feet, directly after which a dense cloud was entered, which brought the temperature down to 36 degrees. At this elevation the report of a gun was heard. Here Mr. Glaisher attempted (probably for the first time in history) to take a cloud-scape photograph, the illumination being brilliant, and the plates with which he was furnished being considered extremely sensitive. The attempt, however, was unsuccessful. The height of two miles was reached in 19 minutes, and here the temperature was at freezing point. In six minutes later three miles was reached, and the thermometer was down to 18 degrees. In another twelve minutes four miles was attained, with the thermometer recording 8 degrees, and by further discharge of sand the fifth aerial milestone was passed at 1.50 p.m., i.e. in 47 minutes from the start, with the thermometer 2 degrees below zero.

Mr. Glaisher relates that up to this point he had taken observations with comfort, and experienced no trouble in respiration, whilst Mr.

Coxwell, in consequence of the exertions he had to make, was breathing with difficulty. More sand was now thrown out, and as the balloon rose higher Mr. Glaisher states that he found some difficulty in seeing clearly. But from this point his experiences should be gathered from his own words:—

"About 1.52 p.m., or later, I read the dry bulb thermometer as minus five; after this I could not see the column of mercury in the wet bulb thermometer, nor the hands of the watch, nor the fine divisions on any instrument. I asked Mr. Coxwell to help me to read the instruments. In consequence, however, of the rotatory motion of the balloon, which had continued without ceasing since leaving the earth, the valve line had become entangled, and he had to leave the car and mount into the ring to readjust it. I then looked at the barometer, and found its reading to be 9 3/4 inches, still decreasing fast, implying a height exceeding 29,000 feet. Shortly after, I laid my arm upon the table, possessed of its full vigour; but on being desirous of using it I found it powerless—it must have lost its power momentarily. Trying to move the other arm, I found it powerless also. Then I tried to shake myself, and succeeded, but I seemed to have no limbs. In looking at the barometer my head fell over my left shoulder. I struggled and shook my body again, but could not move my arms. Getting my head upright for an instant only, it fell on my right shoulder; then I fell backwards, my back resting against the side of the car and my head on its edge. In this position my eyes were directed to Mr. Coxwell in the ring. When I shook my body I seemed to have full power over the muscles of the back, and considerably so over those of the neck, but none over either my arms or my legs. As in the case of the arms, so all muscular power was lost in an instant from my back and neck. I dimly saw Mr. Coxwell, and endeavoured to speak, but could not. In an instant intense darkness overcame me, so that the optic nerve lost power suddenly; but I was still conscious, with as active a brain as at the present moment whilst writing this. I thought I had been seized with asphyxia, and believed I should experience nothing more, as death would come unless we speedily descended. Other thoughts were entering my mind when I suddenly became unconscious, as on going to sleep. I cannot tell anything of the sense of hearing, as no sound reaches the ear to break the perfect stillness and silence of the regions between six and seven miles above the earth. My last observation was made at 1.54 p.m., above 29,000 feet. I suppose two or three minutes to have elapsed between my eyes becoming insensible to seeing fine divisions and 1.54 p.m., and then two or three minutes more to have passed till I was insensible, which I think, therefore, took place about 1.56 p.m. or 1.57 p.m.

"Whilst powerless, I heard the words 'Temperature' and 'Observation,' and I knew Mr. Coxwell was in the car speaking to and endeavouring to rouse me—therefore consciousness and hearing had returned. I then heard him speak more emphatically, but could not see, speak, or move. I heard him again say, 'Do try, now do!' Then the instruments became dimly visible, then Mr. Coxwell, and very shortly I saw clearly. Next, I arose in my seat and looked around, as though waking from sleep, though not refreshed, and said to Mr. Coxwell, 'I have been insensible.' He said, 'You have, and I too, very nearly.' I then drew up my legs, which had been extended, and took a pencil in my hand to begin observations. Mr. Coxwell told me that he had lost the use of his hands, which were black, and I poured brandy over them."

Mr. Glaisher considers that he must have been totally insensible for a period of about seven minutes, at the end of which time the water reserved for the wet bulb thermometer, which he had carefully kept from freezing, had become a solid block of ice. Mr. Coxwell's hands had become frostbitten, so that, being in the ring and desirous of coming to his friend's assistance, he was forced to rest his arms on the ring and drop down. Even then, the table being in the way, he was unable to approach, and, feeling insensibility stealing over himself, he became anxious to open the valve. "But in consequence of having lost the use of his hands he could not do this. Ultimately he succeeded by seizing the cord in his teeth and dipping his head two or three times until the balloon took a decided turn downwards." Mr. Glaisher adds that no inconvenience followed his insensibility, and presently dropping in a country where no conveyance of any kind could be obtained, he was able to walk between seven and eight miles.

The interesting question of the actual height attained is thus discussed by Mr. Glaisher: — "I have already said that my last observation was made at a height of 29,000 feet. At this time, 1.54 p.m., we were ascending at the rate of 1,000 feet per minute, and when I resumed observations we were descending at the rate of 2,000 feet per minute. These two positions must be connected, taking into account the interval of time between, namely, thirteen minutes; and on these considerations the balloon must have attained the altitude of 36,000 or 37,000 feet. Again, a very delicate minimum thermometer read minus 11.9, and this would give a height of 37,000 feet. Mr. Coxwell, on coming from the ring, noticed that the centre of the aneroid barometer, its blue hand, and a rope attached to the car, were all in the same straight line, and this gave a reading of seven inches, and leads to the same result. Therefore, these independent means all lead to about the same elevation, namely, fully seven miles."

So far we have followed Mr. Glaisher's account only, but Mr. Coxwell has added testimony of his own to this remarkable adventure, which renders the narrative more complete. He speaks of the continued rotation of the balloon and the necessity for mounting into the ring to get possession of the valve line. "I had previously," he adds, "taken off a thick pair of gloves so as to be the better able to manipulate the sand-bags, and the moment my unprotected hands rested on the ring, which retained the temperature of the air, I found that they were frost-bitten; but I did manage to bring down with me the valve line, after noticing the hand of the aneroid barometer, and it was not long before I succeeded in opening the shutters in the way described by Mr. Glaisher.... Again, on letting off more gas, I perceived that the lower part of the balloon was rapidly shrinking, and I heard a sighing, as if it were in the network and the ruffled surface of the cloth. I then looked round, although it seemed advisable to let off more gas, to see if I could in any way assist Mr. Glaisher, but the table of instruments blocked the way, and I could not, with disabled hands, pass beneath. My last hope, then, was in seeking the restorative effects of a warmer stratum of atmosphere.... Again I tugged at the valve line, taking stock, meanwhile, of the reserve ballast in store, and this, happily, was ample.

"Never shall I forget those painful moments of doubt and suspense as to Mr. Glaisher's fate, when no response came to my questions. I began to fear that he would never take any more readings. I could feel the reviving effects of a warmer temperature, and wondered that no signs of animation were noticeable. The hand of the aneroid that I had looked at was fast moving, while the under part of the balloon had risen high above the car. I had looked towards the earth, and felt the rush of air as it passed upwards, but was still in despair when Mr. Glaisher gasped with a sigh, and the next moment he drew himself up and looked at me rather in confusion, and said he had been insensible, but did not seem to have any clear idea of how long until he caught up his pencil and noted the time and the reading of the instruments."

The descent, which was at first very rapid, was effected without difficulty at Cold Weston.

CHAPTER XV. FURTHER SCIENTIFIC VOYAGES OF GLAISHER AND COXWELL.

Early in the following spring we find the same two aeronauts going aloft again on a scientific excursion which had a termination nearly as sensational as the last. The ascent was from the Crystal Palace, and the intention being to make a very early start the balloon for this purpose had been partially filled overnight; but by the morning the wind blew strongly, and, though the ground current would have carried the voyagers in comparative safety to the southwest, several pilots which were dismissed became, at no great height, carried away due south. On this account the start was delayed till 1 p.m., by which time the sky had nearly filled in, with only occasional gleams of sun between the clouds. It seemed as if the travellers would have to face the chance of crossing the Channel, and while, already in the car, they were actually discussing this point, their restraining rope broke, and they were launched unceremoniously into the skies. This occasioned an unexpected lurch to the car, which threw Mr. Glaisher among his instruments, to the immediate destruction of some of them.

Another result of this abrupt departure was a very rapid rise, which took the balloon a height of

3,000 feet in three minutes' space, and another 4,000 feet higher in six minutes more. Seven thousand feet vertically in nine minutes is fast pace; but the voyagers were to know higher speed yet that day when the vertical motion was to be in the reverse and wrong direction. At the height now reached they were in cloud, and while thus enveloped the temperature, as often happens, remained practically stationary at about 32 degrees, while that of the dew point increased several degrees. But, on passing out of the cloud, the two temperatures were very suddenly separated, the latter decreasing rapidly under a deep blue upper sky that was now without a cloud. Shortly after this the temperature dropped suddenly some 8 degrees, and then, during the next 12,000 feet, crept slowly down by small stages. Presently the balloon, reaching more than twenty thousand feet, or, roughly, four miles, and still ascending, the thermometer was taken with small fits of rising and falling alternately till an altitude of 24,000 feet was recorded, at which point other and more serious matters intruded themselves.

The earth had been for a considerable time lost to view, and the rate and direction of recent progress had become merely conjectural. What might be taking place in these obscured and lofty regions? It would be as well to discover. So the valve was opened rather freely, with the result that the balloon dropped a mile in three minutes. Then another mile slower, by a shade. Then at 12,000 feet a cloud layer was reached, and shortly after the voyagers broke through into the clear below.

At that moment Mr. Glaisher, who was busy with his instruments, heard Mr. Coxwell make an exclamation which caused him to look over the car, and he writes, "The sea seemed to be under us. Mr. Coxwell again exclaimed, 'There's not a moment to spare: we must save the land at all risks. Leave the instruments.' Mr. Coxwell almost hung to the valve line, and told me to do the same, and not to mind its cutting my hand. It was a bold decision opening the valve in this way, and it was boldly carried out." As may be supposed, the bold decision ended with a crash. The whole time of descending the four and a quarter miles was a quarter of an hour, the last two miles taking four minutes only. For all that, there was no penalty beyond a few bruises and the wrecking of the instruments, and when land was reached there was no rebound; the balloon simply lay inert hard by the margin of the sea. This terrific experience in its salient details is strangely similar to that already recorded by Albert Smith.

In further experimental labours conducted during the summer of this year, many interesting facts stand out prominently among a voluminous mass of observations. In an ascent in an east wind from the Crystal Palace in early July it was found that the upper limit of that wind was reached at 2,400 feet, at which level an air-stream from the north was encountered; but at 3,000 feet higher the wind again changed to a current from the N.N.W. At the height, then, of little more than half a mile, these upper currents were travelling leisurely; but what was more noteworthy was their humidity, which greatly increased with altitude, and a fact which may often be noted here obtruded itself, namely, when the aeronauts were at the upperlimits of the east wind, flat-bottomed cumulus clouds were floating at their level. These clouds were entirely within the influence of the upper or north wind, so that their under sides were in contact with the east wind, i.e. with a much drier air, which at once dissipated all vapour in contact with it, and thus presented the appearance of flat-bottomed clouds. It is a common experience to find the lower surface of a cloud mowed off flat by an east wind blowing beneath it.

At the end of June a voyage from Wolverton was accomplished, which yielded remarkable results of much real value and interest. The previous night had been perfectly calm, and through nearly the whole morning the sun shone in a clear blue sky, without a symptom of wind or coming change. Shortly before noon, however, clouds appeared aloft, and the sky assumed an altered aspect. Then the state of things quickly changed. Wind currents reached the earth blowing strongly, and the half-filled balloon began to lurch to such an extent that the inflation could only with difficulty be proceeded with. Fifty men were unable to hold it in sufficient restraint to prevent rude bumping of the car on the ground, and when, at length, arrangements were complete and release effected, rapid discharge of ballast alone saved collision with neighbouring buildings.

It was now that the disturbance overhead came under investigation; and, considering the short period it had been in progress, proved most remarkable, the more so the further it was explored. At 4,000 feet they plunged into the cloud canopy, through which as it was painfully cold, they, sought to penetrate into the clear above, feeling confident of finding themselves, according to their usual experience, in bright blue sky, with the sun brilliantly shining. On the contrary, however, the region they now entered was further obscured with another canopy of cloud far up. It was while they were traversing this clear interval that a sound unwonted in balloon travel assailed their ears. This was the "sighing, or rather moaning, of the wind as preceding a storm." Rustling of the silk within the cordage is often heard aloft, being due to expansion of gas or similar cause; but the aeronauts soon convinced themselves that what they heard was attributable to nothing else than the actual conflict of air currents beneath. Then they reached fog—a dry fog—and, passing through it, entered a further fog, but wetting this time, and within the next 1,000 feet they were once again in fog that was dry; and then, reaching three miles high and seeing struggling sunbeams, they looked around and saw cloud everywhere, below, above, and far clouds on their own level. The whole sky had filled in most completely since the hours but recently passed, when they had been expatiating on the perfect serenity of the empty heavens.

Still they climbed upwards, and in the next 2,000 feet had entered further fog, dry at first, but turning wetter as they rose. At four miles high they found themselves on a level with clouds, whose dark masses and fringed edges proved them to be veritable rain clouds; and, while still observing them, the fog surged up again and shut out the view, and by the time they had surmounted it they were no less than 23,000 feet up, or higher than the loftiest of the Andes. Even here, with cloud masses still piling high overhead, the eager observer, bent on further quests, was for pursuing the voyage; but Mr. Coxwell interposed with an emphatic, "Too short of sand!" and the downward journey had to be commenced. Then phenomena similar to those already described were experienced again—fog banks (sometimes wet, sometimes dry), rain showers, and cloud strata of piercing cold. Presently, too, a new wonder for a midsummer afternoon—a snow scene all around, and spicules of ice settling and remaining frozen on the coatsleeve. Finally dropping to earth helplessly through the last 5,000 feet, with all ballast spent, Ely Cathedral was passed at close quarters; yet even that vast pile was hidden in the gloom that now lay over all the land.

It was just a month later, and day broke with thoroughly dirty weather, a heavy sky, and falling showers. This was the day of all others that Mr. Glaisher was waiting for, having determined on making special investigations concerning the formation of rain in the clouds themselves. It had long been noticed that, in an ordinary way, if there be two rain gauges placed, one near the surface of the ground, and another at a somewhat higher elevation, then the lower gauge will collect most water. Does, then, rain condense in some appreciable quantity out of the lowest level? Again, during rain, is the air saturated completely, and what regulates the quality of rainfall, for rain sometimes falls in large drops and sometimes in minute particles? These were questions which Mr. Glaisher sought to solve, and there was another.

Charles Green had stated as his conviction that whenever rain was falling from an overcast sky there would always be found a higher canopy of cloud over-hanging the lower stratum. On the day, then, which we are now describing, Mr. Glaisher wished to put this his theory to the test; and, if correct, then he desired to measure the space between the cloud layers, to gauge their thickness, and to see if above the second stratum the sun was shining. The main details of the ascent read thus:—

In ten seconds they were in mist, and in ten seconds more were level with the cloud. At 1,200 feet they were out of the rain, though not yet out of the cloud. Emerging from the lower cloud at 2,300 feet, they saw, what Green would have foretold, an upper stratum of dark cloud above. Then they made excursions up and down, trying

During his researches so far Mr. Glaisher had found much that was anomalous in the way of the winds, and in other elements of weather. He was destined to find much more. It had been commonly accepted that the temperature of the air decreases at the average rate of 10 degrees for every 300 feet of elevation, and various computations, as, for example, those which relate to the co-efficient of refraction, have been founded on this basis; but Mr. Glaisher soon established that the above generalisation had to be much modified. The following, gathered from his notes is a typical example of such surprises as the aeronaut with due instrumental equipment may not unfrequently meet with.

It was the 12th of January, 1864, with an air-current on the ground from the S.E., of temperature 41 degrees,, which very slowly decreased up to 1,600 feet when a warm S.W. current was met with, and at 3,000 feet the temperature was 3 1/2 degrees higher than on the earth. Above the S.W. stream the air became dry, and here the temperature decreased reasonably and consistently with altitude; while fine snow was found falling out of this upper space into the warmer stream below. Mr. Glaisher discusses the peculiarity and formation of this stream in terms which will repay consideration.

"The meeting with this S.W. current is of the highest importance, for it goes far to explain why England possesses a winter temperature so much higher than is due to her northern latitude. Our high winter temperature has hitherto been mostly referred to the influence of the Gulf Stream. Without doubting the influence of this natural agent, it is necessary to add the effect of a parallel atmospheric current to the oceanic current coming from the same region—a true aerial Gulf Stream. This great energetic current meets with no obstruction in coming to us, or to Norway, but passes over the level Atlantic without interruption from mountains. It cannot, however, reach France without crossing Spain and the lofty range of the Pyrenees, and the effect of these cold mountains in reducing its temperature is so great that the former country derives but little warmth from it."

An ascent from Woolwich, arranged as near the equinox of that year as could be managed, supplied some further remarkable results. The temperature, which was 45 degrees to begin with, at 4.7 p.m., crept down fairly steadily till 4,000 feet altitude was registered, when, in a region of warm fog, it commenced rising abruptly, and at 7,500 feet, in blue sky, stood at the same reading as when the balloon had risen only 1,500 feet.

Then, amid many anomalous vicissitudes, the most curious, perhaps, was that recorded late in the afternoon, when, at 10,000 feet, the air was actually warmer than when the ascent began.

That the temperature of the upper air commonly commences to rise after nightfall as the warmth radiated through day hours off the earth collects aloft, is a fact well known to the balloonist, and Mr. Glaisher carried out with considerable success a well-arranged programme for investigating the facts of the case. Starting from Windsor on an afternoon of late May, he so arranged matters that his departure from earth took place about an hour and three quarters before sunset, his intention being to rise to a definite height, and with as uniform a speed as possible to time his descent so as to reach earth at the moment of sundown; and then to re-ascend and descend again m a precisely similar manner during an hour and three-quarters after sunset, taking observations all the way. Ascending for the first flight, he left a temperature of 58 degrees on the earth, and found it 55 degrees at 1,200 feet, then 43 degrees at 3,600 feet, and 29 1/2 degrees at the culminating point of 6,200 feet. Then, during the descent, the temperature increased, though not uniformly, till he was nearly brushing the tops of the trees, where it was some 3 degrees colder than at starting.

It was now that the balloon, showing a little waywardness, slightly upset a portion of the experiment, for, instead of getting to the neighbourhood of earth just at the moment of sunset, the travellers found themselves at that epoch 600 feet above the ground, and over the ridge of a hill, on passing which the balloon became sucked down with a down draught, necessitating a liberal discharge of sand to prevent contact with the ground. This circumstance, slight in itself, caused the lowest point of the descent to be reached some minutes late, and, still more unfortunate, occasioned the ascent which immediately followed to be a rapid one, too rapid, doubtless, to give the registering instruments a fair chance; but one principal record aimed at was obtained at least with sufficient truth, namely, that at the culminating point, which again was 6,200 feet, the temperature read 35 degrees, or about 6 degrees warmer than when the balloon was at the same altitude a little more than an hour before. This comparatively warm temperature was practically maintained for a considerable portion of the descent.

We may summarise the principal of Mr. Glaisher's generalisations thus, using as nearly as possible his own words: —

"The decrease of temperature, with increase of elevation, has a diurnal range, and depends upon the hour of the day, the changes being the greatest at mid-day and the early part of the afternoon, and decreasing to about sunset, when, with a clear sky, there is little or no change of temperature for several hundred feet from the earth; whilst, with a cloudy sky, the change decreases from the mid-day hours at a less rapid rate to about sunset, when the decrease is nearly uniform and at the rate of 1 degree in 2,000 feet.

"Air currents differing in direction are almost always to be met with. The thicknesses of these were found to vary greatly. The direction of the wind on the earth was sometimes that of the whole mass of air up to 20,000 feet nearly, whilst at other times the direction changed within 500 feet of the earth Sometimes directly opposite currents were met with."

With regard to the velocity of upper currents, as shown by the travel of balloons, when the distances between the places of ascent and descent are measured, it was always found that these distances were very much greater than the horizontal movement of the air, as measured by anemometers near the ground.

CHAPTER XVI. SOME FAMOUS FRENCH AERO-NAUTS.

By this period a revival of aeronautics in the land of its birth had fairly set in. Since the last ascents of Gay Lussac, in 1804, already recorded, there had been a lull in ballooning enterprise in France, and no serious scientific expeditions are recorded until the year 1850, when MM. Baral and Bixio undertook some investigations respecting the upper air, which were to deal with its laws of temperature and humidity, with the proportion of carbonic acid present in it, with solar heat at different altitudes, with radiation and the polarisation of light, and certain other interesting enquiries.

The first ascent, made in June from the Paris Observatory, though a lofty one, was attended with so much danger and confusion as to be barren of results. The departure, owing to stormy weather, was hurried and illordered, so that the velocity in rising was excessive, the net constricted the rapidly-swelling globe, and the volumes of out-rushing gas half-suffocated the voyagers. Then a large rent occurred, which caused an alarmingly rapid fall, and the two philosophers were reduced to the necessity of flinging away all they possessed, their instruments only excepted.

The landing, in a vineyard, was happily not attended with disaster, and within a month the same two colleagues attempted a second aerial excursion, again in wet weather.

It would seem as if on this occasion, as on the former one, there was some lack of due management, for the car, suspended at a long distance from the balloon proper, acquired violent oscillations on leaving the ground, and dashing first against a tree, and then against a mast, broke some of the instruments. A little later there occurred a repetition on a minor scale of the aeronauts' previous mishap, for a rent appeared in the silk, though, luckily, so low down in the balloon as to be of small consequence, and eventually an altitude of some 19,000 feet was attained. At one time needles of ice were encountered settling abundantly with a crackling sound upon their notebooks. But the most remarkable observation made during this voyage related to an extraordinary fall of temperature which, as recorded, is without parallel. It took place in a cloud mass, 15,000 feet thick, and amounted to a drop of from 15 degrees to -39 degrees.

In 1867 M. C. Flammarion made a few balloon ascents, ostensibly for scientific research. His account of these, translated by Dr. T. L. Phipson, is edited by Mr. Glaisher, and many of the experiences he relates will be found to contrast with those of others. His physical symptoms alone were remarkable, for on one occasion, at an altitude of apparently little over 10,000 feet, he became unwell being affected with a sensation of drowsiness, palpitation, shortness of breath, and singing in the ears, which, after landing gave place to a "fit of incessant gaping" while he states that in later voyages, at but slightly greater altitudes, his throat and lungs became affected, and he was troubled with presence of blood upon the lips. This draws forth a footnote from Mr. Glaisher, which should be commended to all would-be sky voyagers. It runs thus:—"I have never experienced any of these effects till I had long passed the heights reached by M. Flammarion, and at no elevation was there the presence of blood." However, M. Flammarion adduces, at least, one reassuring fact, which will be read with interest. Once, having, against the entreaties of his friends, ascended with an attack of influenza upon him, he came down to earth again an hour or two afterwards with this troublesome complaint completely cured.

It would seem as if the soil of France supplied the aeronaut with certain phenomena not known in England, one of these apparently being the occasional presence of butterflies hovering round the car when at considerable heights. M. Flammarion mentions more than one occasion when he thus saw them, and found them to be without sense of alarm at the balloon or its passengers. Again, the French observer seems seldom to have detected those opposite airstreams which English balloonists may frequently observe, and have such cause to be wary of. His words, as translated, are: — "It appears to me that two or more currents, flowing in different directions, are very rarely met with as we rise in the air, and when two layers of cloud appear to travel in opposite directions the effect is generally caused by the motion of one layer being more rapid than the other, when the latter appears to be moving in a contrary direction." In continuation of these experiences, he speaks of an occasion when, speeding through the air at the rate of an ordinary express train, he was drawn towards a tempest by a species of attraction.

The French aeronaut's estimate of what constitutes a terrific rate of fall differs somewhat from that of others whose testimony we have been recording. In one descent, falling (without reaching earth, however) a distance of 2,130 feet in two minutes, he describes the earth rising up with frightful rapidity, though, as will be observed, this is not nearly half the speed at which either

Mr. Glaisher or Albert Smith and his companions were precipitated on to bare ground. Very many cases which we have cited go to show that the knowledge of the great elasticity of a well-made wicker car may rob a fall otherwise alarming of its terrors, while the practical certainty that a balloon descending headlong will form itself into a natural parachute, if properly managed, reduces enormously the risk attending any mere impact with earth. It will be allowed by all experienced aeronauts that far worse chances lie in some awkward alighting ground, or in the dragging against dangerous obstacles after the balloon has fallen.

Many of M. Flammarion's experiments are remarkable for their simplicity. Indeed, in some cases he would seem to have applied himself to making trials the result of which could not have been seriously questioned. The following, quoting from Dr. Phipson's translation, will serve as an example:—

"Another mechanical experiment was made in the evening, and renewed next day. I wished to verify Galileo's principle of the independence of simultaneous motions. According to this principle, a body which is allowed to fall from another body in motion participates in the motion of the latter; thus, if we drop a marble from the masthead of a ship, it preserves during its fall the rate of motion of the vessel, and falls at the foot of the mast as if the ship were still. Now, if a body falls from a balloon, does it also follow the motion of the latter, or does it fall directly to the earth in a line which is perpendicular to the point at which we let it fall? In the first case its fall would be described by an oblique line. The latter was found to be the fact, as we proved by letting a bottle fall. During its descent it partakes of the balloon's motion, and until it reaches the earth is always seen perpendicularly below the car."

An interesting phenomenon, relating to the formation of fog was witnessed by M. Flammarion in one of his voyages. He was flying low with a fast wind, and while traversing a forest he noticed here and there patches of light clouds, which, remaining motionless in defiance of the strong wind, continued to hang above the summits of the trees. The explanation of this can hardly be doubtful, being analogous to the formation of a night-cap on a mountain peak where warm moist air-currents become chilled against the cold rock surface, forming, momentarily, a patch of cloud which, though constantly being blown away, is as constantly re-formed, and thus is made to appear as if stationary.

The above instructive phenomenon could hardly have been noticed save by an aeronaut, and the same may be said of the following. Passing in a clear sky over the spot where the Marne flows into the Seine, M. Flammarion notes that the water of the Marne, which, as he says, is as yellow now as it was in the time of Julius Caesar, does not mix with the green water of the Seine, which flows to the left of the current, nor with the blue water of the canal, which flows to the right. Thus, a yellow river was seen flowing between two distinct brooks, green and blue respectively.

Here was optical evidence of the way in which streams of water which actually unite may continue to maintain independent courses. We have seen that the same is true of streams of air, and, where these traverse one another in a copious and complex manner, we find, as will be shown, conditions produced that cause a great deadening of sound; thus, great differences in the travel of sound in the silent upper air can be noticed on different days, and, indeed, in different periods of the same aerial voyage. M. Flammarion bears undeniable testimony to the manner in which the equable condition of the atmosphere attending fog enhances, to the aeronaut, the hearing of sounds from below. But when he gives definite heights as the range limits of definite sounds it must be understood that these ranges will be found to vary greatly according to circumstances. Thus, where it is stated that a man's voice may make itself heard at 3,255 feet, it might be added that sometimes it cannot be heard at a considerably less altitude; and, again, the statement that the whistle of a locomotive rises to near 10,000 feet, and the noise of a railway train to 8,200 feet, should be qualified an additional note to the effect that both may be occasionally heard at distances vastly greater. But perhaps the most curious observation of M. Flammarion respecting sounds aloft relates to that of echo. To his fancy, this had a vague depth, appearing also to rise from the horizon with a curious tone, as if it came from another world. To the writer, on the contrary, and to many fellow observers who have specially experimented with this test of sound, the echo has always appeared to come very much from the right place—the spot nearly immediately below—and if this suggested its coming from another world then the same would have to be said of all echoes generally.

About the same period when M. Flammarion was conducting his early ascents, MM. de Fonvielle and Tissandier embarked on experimental voyages, which deserve some particular notice. Interest in the new revival of the art of aeronautics was manifestly be coming reestablished in France, and though we find enthusiasts more than once bitterly complaining of the lack of financial assistance, still ballooning exhibitions, wherever accomplished, never failed to arouse popular appreciation. But enthusiasm was by no means the universal attitude with which the world regarded aerial enterprise. A remarkable and instructive instance is given to the contrary by M. W. de Fonvielle himself.

He records an original ballooning exploit, organised at Algiers, which one might have supposed would have caused a great sensation, and to which he himself had called public attention in the local journals. The brothers Braguet were to make an ascent from the Mustapha Plain in a small fire balloon heated with burning straw, and this risky performance was successfully carried out by the enterprising aeronauts. But, to the onlooker, the most striking feature of the proceeding was the fact that while the Europeans present regarded the spectacle with curiosity and pleasure, the native Mussulmans did not appear to take the slightest interest in it; "And this," remarked de Fonvielle, "was not the first time that ignorant and fanatic people have been noted as manifesting complete indifference to balloon ascents. After the taking of Cairo, when General

Buonaparte wished to produce an effect upon the inhabitants, he not only made them a speech, but supplemented it with the ascent of a fire balloon. The attempt was a complete failure, for the French alone looked up to the clouds to see what became of the balloon."

In the summer of 1867 an attempt was made to revive the long extinct Aeronautic Company of France, established by De Guyton. The undertaking was worked with considerable energy. Some forty or fifty active recruits were pressed into the service, a suitable captive balloon was obtained, thousands of spectators came to watch the evolutions; and many were found to pay the handsome fee of 100 francs for a short excursion in the air. For all this, the effort was entirely abortive, and the ballooning corps, as such, dropped out of existence.

A little while after this de Fonvielle, on a visit to England, had a most pathetic interview with the veteran Charles Green, who was living in comfortable retirement at Upper Holloway. The grand old man pointed to a well-filled portfolio in the corner of his room, in which, he said, were accounts of all his travels, that would require a lifetime to peruse and put in order. Green then took his visitor to the end of the narrow court, and, opening the door of an outhouse, showed him the old Nassau balloon. "Here is my car," he said, touching it with a kind of solemn respect, "which, like its old pilot, now reposes quietly after a long and active career. Here is the guide rope which I imagined in former years, and which has been found very useful to aeronauts.... Now my life has past and my time has gone by.... Though my hair is white and my body too weak to help you, I can still give you my advice, and you have my hearty wishes for your future."

It was but shortly after this, on March 26, 1870, that Charles Green passed away in the 85th year of his age.

De Fonvielle's colleague, M. Gaston Tissandier, was on one occasion accidentally brought to visit the resting place of the earliest among aeronauts, whose tragic death occurred while Charles Green himself was yet a boy. In a stormy and hazardous descent Tissandier, under the guidance of M. Duruof, landed with difficulty on the sea coast of France, when one of the first to render help was a lightkeeper of the Griz-nez lighthouse, who gave the information that on the other side of the hills, a few hundred yards from the spot where they had landed, was the tomb of Pilatre de Rozier, whose tragical death has been recorded in an early chapter. A visit to the actual locality the next day revealed the fact that a humble stone still marked the spot.

Certain scientific facts and memoranda collected by the talented French aeronaut whom we are following are too interesting to be omitted. In the same journey to which we have just referred the voyagers, when nearly over Calais, were witnesses from their commanding standpoint of a very striking phenomenon of mirage. Looking in the direction of England, the far coast line was hidden by an immense veil of leaden-coloured cloud, and, following this cloud wall upward to detect where it terminated, the travellers saw above it a greenish layer like that of the surface of the sea, on which was detected a little black point suggesting a walnut shell. Fixing their eyes on this black spot, they presently discerned it to be a ship sailing upside down upon an aerial ocean. Soon after, a steamer blowing smoke, and then other vessels, added themselves to the illusory spectacle.

Another wonder detected, equally striking though less uncommon, was of an acoustical nature, the locality this time being over Paris. The height of the balloon at this moment was not great, and, moreover, was diminishing as it settled down. Suddenly there broke in upon the voyagers a sound as of a confused kind of murmur. It was not unlike the distant breaking of waves against a sandy coast, and scarcely less monotonous. It was the noise of Paris that reached them, as soon as they sank to within 2,600 feet of the ground, but it disappeared at once when they threw out just sufficient ballast to rise above that altitude.

It might appear to many that so strange and sudden a shutting out of a vast sound occurring abruptly in the free upper air must have been more imaginary than real, yet the phenomenon is almost precisely similar to one coming within the experience the writer, and vouched for by his son and daughter, as also by Mr. Percival Spencer, all of whom were joint observers at the time, the main point of difference in the two cases being the fact that the "region of silence" was recorded by the French observers as occurring at a somewhat lower level. In both cases there is little doubt that the phenomenon can be referred to a stratum of disturbed or non-homogeneous air, which may have been very far spread, and which is capable of acting as a most opaque sound barrier.

Attention has often been called in these pages to the fact that the action of the sun on an inflated balloon, even when the solar rays may be partially obscured and only operative for a few passing moments, is to give sudden and great buoyancy to the balloon. An admirable opportunity for fairly estimating the dynamic effect of the sun's rays on a silk globe, whose fabric was half translucent, was offered to the French aeronauts when their balloon was spread on the grass under repair, and for this purpose inflated with the circumambient air by means of a simple rotatory fan. The sun coming out, the interior of the globe quickly became suffocating, and it was found that, while the external temperature recorded 77 degrees, that of the interior was in excess of 91 degrees.

CHAPTER XVII. ADVENTURE AND ENTERPRISE.

A balloon which has become famous in history was frequently used in the researches of the French aeronauts mentioned in our last chapter. This was known as "The Giant," the creation of M. Nadar, a progressive and practical aeronaut, who had always entertained ambitious ideas about aerial travel.

M. Nadar had been editor of L'Aeronaut, a French journal devoted to the advancement of aerostation generally. He had also strongly expressed his own views respecting the possibility of constructing air ships that should be subject to control and guidance when winds were blowing. His great contention was that the dirigible air ship would, like a bird, have to be made heavier than the medium in which it was to fly. As he put it, a balloon could never properly become a vessel. It would only be a buoy. In spite of any number of accessories, paddles, wings, fans, sails, it could not possibly prevent the wind from bodily carrying away the whole concern.

After this strong expression of opinion, it may appear somewhat strange that such a bold theoriser should at once have set himself to construct the largest gas balloon on record. Such, however, was the case and the reason urged was not otherwise than plausible. For, seeing that a vast sum of money would be needed to put his theories into practice, M. Nadar conceived the idea of first constructing a balloon so unique and unrivalled that it should compel public attention in a way that no other balloon had done before, and so by popular exhibitions bring to his hand such sums as he required. A proper idea of the scale of this huge machine can be easily gathered. The largest balloons at present exhibited in this country are seldom much in excess of 50,000 cubic feet capacity. Compared with these the "Great Nassau Balloon," built by Charles Green, which has been already sufficiently described, was a true leviathan; while Coxwell's "Mammoth" was larger yet, possessing a content, when fully inflated, of no less than 93,000 cubic feet, and measuring over 55 feet in diameter. This, however, as will be seen, was but a mere pigmy when compared with "The Giant," which, measuring some 74 feet in diameter, possessed the prodigious capacity of 215,000 cubic feet.

But the huge craft possessed another novelty besides that of exceptional size. It was provided with a subsidiary balloon, called the "Compensator," and properly the idea of M. L. Godard, the function of which was to receive any expulsion of gas in ascending, and thus to prevent loss during any voyage. The specification of this really remarkable structure may be taken from M. Nadar's own description. The globe in itself was for greater strength virtually double, consisting of two identical balloons, one within the other, each made of white silk of the finest quality, and costing about 5s. 4d. per yard. No less than 22,000 yards of this silk were required, and the sewing up of the gores was entirely done by hand. The small compensating balloon was constructed to have a capacity of about 3,500 cubic feet, and the whole machine, when fully inflated, was calculated to lift 4 1/2 tons. With this enormous margin of buoyancy, M. Nadar determined on making the car of proportionate and unparalleled dimensions, and of most elaborate design. It contained two floors, of which the upper one was open, the height of all being nearly 7 feet, with a width of about 13 feet. Then what was thought to be due provision was made for possible emergencies. It might descend far from help or habitations, therefore means were provided for attaching wheels and axles. Again, the chance of rough impact had to be considered, and so canes, to act as springs, were fitted around and below. Once again, there was the contingency of immersion to be reckoned with; therefore there were provided buoys and water-tight compartments. Further than this, unusual luxuries were added, for there were cabins, one for the captain at one end, and another with three berths for passengers at the other. Nor was this all, for there was, in addition, a larder, a lavatory, a photographic room, and a printing office. It remains now only to tell the tale of how this leviathan of the air acquitted itself.

The first ascent was made on the 4th of October, 1853, from the Champ de Mars, and no fewer than fifteen living souls were launched together into the sky. Of these Nadar was captain, with the brothers Godard lieutenants. There was the Prince de Sayn-Wittgenstein; there was the Count de St. Martin; above all, there was a lady, the Princess de la Tour d'Auvergne. The balloon came to earth at 9 o'clock at night near Meaux, and, considering all the provision which had been made to guard against rough landing, it can hardly be said that the descent was a happy one. It appears that the car dragged on its side for nearly a mile, and the passengers, far from finding security in the seclusion of the inner chambers, were glad to clamber out above and cling, as best they might, to the ropes.

Many of the party were bruised more or less severely, though no one was seriously injured, and it was reported that such fragile articles as crockery, cakes, confectionery, and wine bottles to the number of no less than thirty-seven, were afterwards discovered to be intact, and received due attention. It is further stated that the descent was decided on contrary to the wishes of the captain, but in deference to the judgment of the experienced MM. Godard, it being apparently their conviction that the balloon was heading out to sea, whereas, in reality, they were going due east, "with no sea at all before them nearer than the Caspian."

This was certainly an unpropitious trial trip for the vessel that had so ambitiously sought dominion over the air, and the next trial, which was embarked upon a fortnight later, Sunday, October 18th, was hardly less unfortunate. Again the ascent was from the Champ de Mars, and the send-off lacked nothing in the way of splendour and circumstance. The Emperor was present, for two hours an interested observer of the proceedings; the King of Greece also attended, and even entered the car, while another famous spectator was the popular Meyerbeer. "The Giant" first gave a preliminary demonstration of his power by taking up, for a cable's length, a living freight of some thirty individuals, and then, at 5.10 p.m., started on its second free voyage, with nine souls on board, among them again being a lady, in the person of Madame Nadar. For nearly twenty-four hours no tidings of the voyage were forthcoming, when a telegram was received stating that the balloon had passed over Compiegne, more than seventy miles from Paris, at 8.30 on the previous evening, and that Nadar had dropped the simple message, "All goes well!" A later telegram the same evening stated that the balloon had at midnight on Sunday passed the Belgian frontier over Erquelines, where the Custom House officials had challenged the travellers without receiving an answer.

But eight-and-forty hours since the start went by without further news, and excitement in Paris grew intense. When the news came at last it was from Bremen, to say that Nadar's balloon had descended at Eystrup, Hanover, with five of the passengers injured, three seriously. These three were M. Nadar, his wife, and M. St. Felix. M. Nadar, in communicating this intelligence, added, "We owe our lives to the courage of Jules Godard." The following signed testimony of M. Louis Godard is forthcoming, and as it refers to an occasion which is among the most thrilling in aerial adventure, it may well be given without abridgment. It is here transcribed almost literatim from Mr. H. Turner's valuable work, "Astra Castra."

"The Giant," after passing Lisle, proceeded in the direction of Belgium, where a fresh current, coming from the Channel, drove it over the marshes of Holland. It was there that M. Louis Godard proposed to descend to await the break of day, in order to recognise the situation and again to depart. It was one in the morning, the night was dark, but the weather calm. Unfortunately, this advice, supported by long experience, was not listened to. "The Giant" went on its way, and then Louis Godard no longer considered himself responsible for the consequences of the voyage.

The balloon coasted the Zuyder Zee, and then entered Hanover. The sun began to appear, drying the netting and sides of the balloon, wet from its passage through the clouds, and produced a dilatation which elevated the aeronauts to 15,000 feet. At eight o'clock the wind, blowing suddenly from the west, drove the balloon in a right line towards the North Sea. It was necessary, at all hazards, to effect a descent. This was a perilous affair, as the wind was blowing with extreme violence. The brothers Godard assisted, by M. Gabriel, opened the valve and got out the anchors; but, unfortunately, the horizontal progress of the balloon augmented from second to second. The first obstacle which the anchors encountered was a tree; it was instantly uprooted, and dragged along to a second obstacle, a house, whose roof was carried off. At this moment the two cables of the anchors were broken without the voyagers being aware of it. Foreseeing the successive shocks that were about to ensue—the moment was critical—the least forgetfulness might cause death. To add to the difficulty, the balloon's inclined position did not permit of operating the valve, except on the hoop.

At the request of his brother, Jules Godard attempted the difficult work of climbing to this hoop, and, in spite of his known agility, he was obliged several times to renew the effort. Alone, and not being able to detach the cord, M. Louis Godard begged M. Yon to join his brother on the hoop. The two made themselves masters of the rope, which they passed to Louis Godard. The latter secured it firmly, in spite of the shocks he received. A violent impact shook the car and M. de St. Felix became entangled under the car as it was ploughing the ground. It was impossible to render him any assistance; notwithstanding, Jules Godard, stimulated by his brother, leapt out to attempt mooring the balloon to the trees by means of the ropes. M. Montgolfier, entangled in the same manner, was re-seated in time and saved by Louis Godard.

At this moment others leapt out and escaped with a few contusions. The car, dragged along by the balloon, broke trees more than half a yard in diameter and overthrew everything that opposed it.

Louis Godard made M. Yon leap out of the car to assist Madame Nadar; but a terrible shock threw out MM. Nadar, Louis Godard, and Montgolfier, the two first against the ground, the third into the water. Madame Nadar, in spite of the efforts of the voyagers, remained the last, and found herself squeezed between the ground and the car, which had fallen upon her. More than twenty minutes elapsed before it was possible to disentangle her, in spite of the most vigorous efforts on the part of everyone. It was at this moment the balloon burst and, like a furious monster, destroyed everything around it. Immediately afterwards they ran to the assistance of M. de St. Felix, who had been left behind, and whose face was one ghastly wound, and covered with blood and mire. He had an arm broken, his chest grazed and bruised.

After this accident, though a creditable future lay in store for "The Giant," its monstrous and unwieldy car was condemned, and presently removed to the Crystal Palace, where it was daily visited by large crowds.

It is impossible to dismiss this brief sketch of French balloonists of this period without paying some due tribute to M. Depuis Delcourt, equally well known in the literary and scientific world, and regarded in his own country as a father among aeronauts. Born in 1802, his recollection went back to the time of Montgolfier and Charles, to the feats of Garnerin, and the death of Madame Blanchard. He established the Aerostatic and Meteorological Society of France, and was the author of many works, as well as of a journal dealing with aerial navigation. He closed a life devoted to the pursuit and advancement of aerostation in April, 1864.

Before very long, events began shaping themselves in the political world which were destined to bring the balloon in France into yet greater prominence. But we should mention that already its capabilities in time of war to meet the requirements of military operations had been scientifically and systematically tested, and of these trials it will be necessary to speak without further delay.

Reference has already been made in these pages to a valuable article contributed in 1862 by Lieutenant G. Grover, R.E., to the Royal Engineers' papers. From this report it would appear that the balloon, as a means of reconnoitring, was employed with somewhat uncertain success at the battle of Solferino, the brothers Godard being engaged as aeronauts. The balloon used was a Montgolfier, or fire balloon, and, in spite of its ready inflation, MM. Godard considered it, from the difficulty of maintaining within it the necessary degree of buoyancy, far inferior to the gas inflated balloon. On the other hand, the Austrian Engineer Committee were of a contrary opinion. It would seem that no very definite conclusions had been arrived at with respect to the use and value of the military balloon up to the time of the commencement of the American War in 1862.

It was now that the practice of ballooning became a recognised department of military manoeuvres, and a valuable report appears in the above-mentioned papers from the pen of Captain F. Beaumont, R.E. According to this officer, the Americans made trial of two different balloons, both hydrogen inflated, one having a capacity of about 13,000 cubic feet, and the other about twice as large. It was this latter that the Americans used almost exclusively, it being found to afford more steadiness and safety, and to be the means, sometimes desirable, of taking up more than two persons. The difficulty of sufficient gas supply seems to have been well met. Two generators sufficed, these being "nothing more than large tanks of wood, acid-proof inside, and of sufficient strength to resist the expansive action of the gas; they were provided with suitable stopcocks for regulating the admission of the gas, and with manhole covers for introducing the necessary materials." The gas, as evolved, being made to pass successively through two vessels containing lime water, was delivered cool and purified into the balloon, and as the sulphuric acid needed for the process was found sufficiently cheap, and scrap iron also required was readily come by, it would seem that practical difficulties in the field were reduced to a minimum.

According to Captain Beaumont, the difficulties which might have been expected from windy weather were not considerable, and twenty-five or thirty men sufficed to convey the balloon easily, when inflated, over all obstacles. The transport of the bulk of the rest of the apparatus does not read, on paper, a very serious matter. The two generators required four horses each, and the acid and balloon carts as many more. Arrived on the scene of action, the drill itself was a simple matter. A squad of thirty men under an officer sufficed to get the balloon into position, and to arrange the ballast so that, with all in, there was a lifting power of some thirty pounds. Then, at the word of command, the men together drop the car, and seize the three guy ropes, of which one is made to pass through a snatch block firmly secured. The guy ropes are then payed out according to the directions of the aeronaut, as conveyed through the officer.

The balloon accompanied the army's advance where its services could be turned to the greatest advantage. It was employed in making continual ascents, and furnishing daily reports to General M'Clellan, and it was supposed that by constant observation the aeronaut could, at a glance, assure himself that no change had taken place in the occupation of the country. Captain Beaumont, speaking, be it remembered, of the military operations and manoeuvres then in vogue, declared that earthworks could be seen even at the distance of eight miles, though their character could not be distinctly stated. Wooded country was unfitted for balloon reconnaissance, and only in a plain could any considerable body of troops be made known. Then follows such a description as one would be expecting to find:—

"During the battle of Hanover Court House, which was the first engagement of importance before Richmond, I happened to be close to the balloon when the heavy firing began. The wind was rather high; but I was anxious to see, if possible, what was going on, and I went up with the father of the aeronaut. The balloon was, however, short of gas, and as the wind was high we were obliged to come down. I then went up by myself, the diminished weight giving increased steadiness; but it was not considered safe to go more than 500 feet, on account of the unsettled state of the weather. The balloon was very unsteady, so much so that it was difficult to fix my sight on any particular object. At that distance I could see nothing of the fight."

Following this is another significant sentence: —

"In the case of a siege, I am inclined to think that a balloon reconnaissance would be of less value than in almost any other case where a reconnaissance can be required; but, even here, if useless, it is, at any rate, also harmless. I once saw the fire of artillery directed from the balloon; this became necessary, as it was only in this way that the picket which it was desired to dislodge could be seen. However, I cannot say that I thought the fire of artillery was of much effect against the unseen object; not that this was the fault of the balloon, for had it not told the artillerists which way the shots were falling their fire would have been more useless still."

It will be observed that at this time photography had not been adopted as an adjunct to military ballooning.

Full details have been given in this chapter of the monster balloon constructed by M. Nadar; but in 1864 Eugene Godard built one larger yet of the Montgolfier type. Its capacity was nearly half a million cubic feet, while the stove which inflated it stood 18 feet high, and weighed nearly 1,000 pounds. Two free ascents were made without mishap from Cremorne Gardens. Five years later Ashburnham Park was the scene of captive ascents made with another mammoth balloon, containing no less than 350,000 cubic feet of pure hydrogen, and capable of lifting 11 tons. It was built at a cost of 28,000 francs by M. Giffard, the well-known engineer and inventor of the injector for feeding steam engines.

These aerial leviathans do not appear to have been, in any true sense successful.

CHAPTER XVIII. THE BALLOON IN THE SIEGE OF PARIS.

Within a few months of the completion of the period covered by the records of the last chapter, France was destined to receive a more urgent stimulus than ever before to develop the resources of ballooning, and, in hot haste, to turn to the most serious and practical account all the best resources of aerial locomotion. The stern necessity of war was upon her, and during four months the sole mode of exit from Paris—nay, the only possible means of conveying a simple message beyond the boundary of her fortifications—was by balloon.

Hitherto, from the very inception of the art from the earliest Montgolfier with its blazing furnace, the balloon had gone up from the gay capital under every variety of circumstance—for pleasure, for exhibition, for scientific research. It was now put in requisition to mitigate the emergency occasioned by the long and close investment of the city by the Prussian forces.

Recognising, at an early stage, the possibilities of the balloon, an enquiry was at once made by the military authorities as to the existing resources of the city, when it was quickly discovered that, with certain exceptions to be presently mentioned, such balloons as were in existence within the walls were either unserviceable or inadequate for the work that was demanded of them. Thereupon, with admirable promptness and enterprise, it was forthwith determined to organise the building and equipment of a regular fleet of balloons of sufficient size and strength.

It chanced that there were in Paris at the time two professional aeronauts of proved experience and skill, both of whom had become well known in London only the season before in connection with M. Giffard's huge captive balloon at Ashburnham Park. These were MM. Godard and Yon, and to them was entrusted the establishment of two separate factories in spacious buildings, which were at once available and admirably adapted for the purpose. These were at the Orleans and the Northern Railway stations respectively, where spacious roofs and abundant elbow room, the chief requisites, were to be found. The first-mentioned station was presided over Godard, the latter by M. Yon, assisted by M. Dartois.

It was not doubted that the resources of the city would be able to supply the large demand that would be made for suitable material; but silk as a fabric was at once barred on the score of expense alone. A single journey was all that needed to be calculated on for each craft, and thus calico would serve the purpose, and would admit of speedy making up. Slight differences in manufacture were adopted at the two factories. At the Northern station plain white calico was used, sewn with a sewing machine, whereas at the Orleans station the material was coloured and entrusted only to hand stitching. The allimportant detail of varnish was supplied by a mixture of linseed oil and the active principle of ordinary driers, and this, laid on with a rubber, rendered the material gas-tight and quickly dry enough for use. Hundreds of hands, men and women, were employed at the two factories, at which some sixty balloons were produced before the end of the siege. Much of the more important work was entrusted to sailors, who showed special aptness, not only in fitting out and rigging the balloons, but also in their management when entrusted to the winds.

It must have been an impressive sight for friend or foe to witness the departure of each aerial vessel on its venturesome mission. The bold plunge into space above the roofs of the imprisoned city; the rapid climb into the sky and, later, the pearl drop high in air floating away to its uncertain and hazardous haven, running the gauntlet of the enemy's fire by day or braving what at first appeared to be equal danger, attending the darkness of night. It will be seen, however, that, of the two evils, that of the darkness was considered the less, even though, with strange and unreasonable excess of caution, the aeronauts would not suffer the use of the perfectly safe and almost indispensable Davy lamp.

Before any free ascents were ventured on, two old balloons were put to some practical trial as stationary observatories. One of these was moored at Montmartre, the other at Mont-souris. From these centres daily, when the weather permitted, captive ascents were made—four by day and two by night—to watch and locate the movements of the enemy. The system, as far as it went, was well planned. It was safe, and, to favour expedition, messages were written in the car of the balloon and slid down the cable to the attendants below. The net result, however, from a strategic point of view, does not appear to have been of great value.

Ere yet the balloons were ready, certain bold and eventful escapes were ventured on. M. Duruof, already introduced in these pages, trusting himself to the old craft, "Le Neptune," in unskyworthy condition, made a fast plunge into space, and, catching the upper winds, was borne away for as long a period as could be maintained at the cost of a prodigal expenditure of ballast.

The balloon is said to have described a visible parabola, like the trajectory of a projectile, and fell at Evreux in safety and beyond the range of the enemy's fire, though not far from their lines. This was on the 23rd of September. Two days afterwards the first practical trial was made with homing pigeons, with the idea of using them in connection with balloons for the establishment of an officially sanctioned post. MM. Maugin and Grandchamp conducted this voyage in the "Ville de Florence," and descended near Vernouillet, not far beyond Le Foret de St. Germain, and less than twenty miles from Paris. The serviceability of the pigeon, however, was clearly established, and a note contributed by Mr. Glaisher, relating to the breeding and choice of these birds, may be considered of interest. Mr. R. W. Aldridge, of Charlton, as quoted by Mr. Glaisher, stated that his experience went to show that these birds can be produced with different powers of orientation to meet the requirements of particular cases. "The bird required to make journeys under fifty miles would materially differ in its pedigree from one capable of flying 100 or 600 miles. Attention, in particular, must be given to the colour of the eye; if wanted for broad daylight the bird known as the 'Pearl Eye,' from its colour, should be selected; but if for foggy weather or for twilight flying high and low to verify these conditions, and passing through fogs both wet and dry, at last drifting earthward, through squalls of wind and rain with drops as large as fourpenny pieces, to find that on the ground heavy wet had been ceaselessly falling.

A day trip over the eastern suburbs of London in the same year seems greatly to have impressed Mr. Glaisher. The noise of London streets as heard from above has much diminished during the last fifteen years' probably owing to the introduction of wood paving. But, forty years ago, Mr. Glaisher describes the deep sound of London as resembling the roar of the sea, when at a mile high; while at greater elevations it was heard at a murmuring noise. But the view must have been yet more striking than the hearing, for in one direction the white cliffs from Margate to Dover were visible, while Brighton and the sea beyond were sighted, and again all the coast line up to Yarmouth yet the atmosphere that day, one might have thought, should have been in turmoil, by reason of a conflict of aircurrents; for, within two miles of the earth, the wind was from the east; between two and three miles high it was exactly opposite, being from the west; but at three miles it was N.E.; while, higher, it was again directly opposite, or S.W.

the black- or blue-eyed bird should receive the preference."

Only a small minority, amounting to about sixty out of 360 birds taken up, returned to Paris, but these are calculated to have conveyed among them some 100,000 messages. To reduce these pigeon messages to the smallest possible compass a method of reduction by photography was employed with much success. A long letter might, in this way, be faithfully recorded on a surface of thinnest photographic paper, not exceeding the dimensions of a postage stamp, and, when received, no more was necessary than to subject it to magnification, and then to transcribe it and send a fair copy to the addressee.

The third voyage from Paris, on September 29th was undertaken by Louis Godard in two small balloons, united together, carrying both despatches and pigeons, and a safe landing was effected at Mantes This successful feat was rival led the next day by M. Tissandier, who ascended alone in a balloon of only some 26,000 cubic feet capacity and reached earth at Dreux, in Normandy.

These voyages exhausted the store of ready-made balloons, but by a week later the first of those being specially manufactured was ready, and conveyed in safety from the city no less a personage than M. Gambetta.

The courageous resolve of the great man caused much sensation in Paris, the more so because, owing to contrary winds, the departure had to be postponed from day to day. And when, at length, on October 7th, Gambetta and his secretary, with the aeronaut Trichet, actually got away, in company with another balloon, they were vigorously fired at with shot and shell before they had cleared St. Denis. Farther out over the German posts they were again under fire, and escaped by discharging ballast, not, however, before Gambetta had been grazed by a bullet. Yet once more they were assailed by German volleys before, about 3 p.m., they found a haven near Montdidier.

The usual dimensions of the new balloons gave a capacity of 70,000 cubic feet, and each of these, when inflated with coal gas, was calculated to convey a freight of passengers, ballast, and despatches amounting to some 2,000 pounds. Their despatch became frequent, sometimes two in the same twenty-four hours. In less than a single week in October as many as four balloons had fallen in Belgium, and as many more elsewhere. Up till now some sixteen ventures had ended well, but presently there came trouble. On October 22nd MM. Iglesia and Jouvencel fell at Meaux, occupied by the Prussians; their despatches, however, were saved in a dung cart. The twenty-third voyage ended more unhappily. On this occasion a sailor acted as aeronaut, accompanied by an engineer, Etienne Antonin, and carrying nearly 1,000 pounds of letters. It chanced that they descended near Orleans on the very day when that town was re-occupied by the enemy, and both voyagers were made prisoners. The engineer, however, subsequently escaped. Three days later another sailor, also accompanied by an engineer, fell at the town of Ferrieres, then occupied by the Prussians, when both were made prisoners. In this case, also, the engineer succeeded in making his escape; while the despatches were rescued by a forester and forwarded in safety.

At about this date W. de Fonvielle, acting as aeronaut, and taking passengers, made a successful escape, of which he has given a graphic account. He had been baulked by more than one serious contretemps. It had been determined that the departure should be by night, and November 19th being fixed upon, the balloon was in process of inflation under a gentle wind that threatened a travel towards Prussian soil, when, as the moment of departure approached, a large hole was accidentally made in the fabric by the end of the metal pipe, and it was then too late to effect repairs. The next and following days the weather was foul, and the departure was not effected till the 25th, when he sailed away over the familiar but desolated country. He and his companions were fired at, but only when they were well beyond range, and in less than two hours the party reached Louvain, beyond Brussels, some 180 English miles in a direct line from their starting point. This was the day after the "Ville d'Orleans" balloon had made the record voyage and distance of all the siege, falling in Norway, 600 miles north of Christiania, after a flight of fifteen hours.

At the end of November, when over thirty escape voyages had been made, two fatal disasters occurred. A sailor of the name of Prince ascended alone on a moonless night, and at dawn, away on the north coast of Scotland, some fishermen sighted a balloon in the sky dropping to the westward in the ocean. The only subsequent trace of this balloon was a bag of despatches picked up in the Channel. Curiously enough, two days later almost the same story was repeated. Two aeronauts, this time in charge of despatches and pigeons, were carried out to sea and never traced.

Undeterred by these disasters, a notable escape was now attempted. An important total eclipse of the sun was to occur in a track crossing southern Spain and Algeria on December 22nd. An enthusiastic astronomer, Janssen, was commissioned by the Academy of Sciences to attend and make observations of this eclipse. But M. Janssen was in Paris, as were also his instruments, and the eclipse track lay nearly a thousand miles away. The one and only possible mode of fulfilling his commission was to try the off-chance afforded by balloon, and this chance he resorted to only twenty days before the eclipse was due.

Taking with him the essential parts of a reflecting telescope, and an active young sailor as assistant, he left Paris at 6 a.m. and rose at once to 3,600 feet, dipping again somewhat at sunrise (owing, as he supposed, to loss of heat through radiation), but subsequently ascending again rapidly under the increased altitude of the sun till his balloon attained its highest level of 7,200 feet. From this elevation, shortly after 11 a.m., he sighted the sea, when he commenced a descent which brought him to earth at the mouth of the Loire. It had been fast travelling—some 300 miles in little more than three hours—and the ground wind was strong. Nevertheless, neither passengers nor instruments were injured, and M. Janssen was fully established by the day of eclipse on his observing ground at Oran, on the

Algerian coast. It is distressing to add that the phenomenon was hidden by cloud. In the month that followed this splendid venture no fewer than fifteen balloons escaped from Paris, of which four fell into the hands of the enemy, although for greater security all ascents were now being made by night.

On January 13th, 1871, a new device for the return post was tried, and, in addition to pigeons, sheep dogs were taken up, with the idea of their being returned to the city with messages concealed within their collars. There is apparently no record of any message having been returned to the town by this ingenious method. On January 24th a balloon, piloted by a sailor, and containing a large freight of letters, fell within the Prussian lines, but the patriotism of the country was strong enough to secure the despatches being saved and entrusted to the safe conveyance of the Post Office. Then followed the total loss of a balloon at sea; but this was destined to be the last, save one, that was to attempt the dangerous mission. The next day, January 28th, the last official balloon left the town, manned by a single sailor, carrying but a small weight of despatches, but ordering the ships to proceed to Dieppe for the revictualling of Paris.

Five additional balloons at that time in readiness were never required for the risky service for which they were designed.

There can be little doubt that had the siege continued a more elaborate use of balloons would have been developed. Schemes were being mooted to attempt the vastly more difficult task of conveying balloons into Paris from outside. When hostilities terminated there were actually six balloons in readiness for this venture at Lisle, and waiting only for a northerly wind. M. de Fonvielle, possessed of both courage and experience, was prepared to put in practice a method of guiding by a small propelling force a balloon that was being carried by sufficiently favouring winds within a few degrees of its desired goal — and in the case of Paris the goal was an area of some twenty miles in diameter. Within the invested area several attempts were actually made to control balloons by methods of steering. The names of Vert and Dupuy de Lome must here be specially mentioned. The former had elaborated an invention which received much assistance, and was subsequently exhibited at the Crystal Palace. The latter received a grant of L1,600 to perfect a complex machine, having within its gas envelope an air chamber, suggested by the swimming bladder of a fish, having also a sail helm and a propelling screw, to be operated by manual labour.

The relation of this invention to others of similar purpose will be further discussed later on. But an actual trial of a dirigible craft, the design of Admiral Labrousse, was made from the Orleans railway station on January 9th. This machine consisted of a balloon of about the standard capacity of the siege balloons, namely some 70,000 cubic feet, fitted with two screws of about 12 feet diameter, but capable of being readily worked at moderate speed. It was not a success. M. Richard, with three sailors, made a tentative ascent, and used their best endeavours to control their vessel, but practically without avail, and the machine presently coming to earth clumsily, a portion of the gear caught in the ground and the travellers were thrown over and roughly dragged for a long distance.

Fairly looked at, the aerial post of the siege of Paris must be regarded as an ambitious and, on the whole, successful enterprise. Some two million and a half of letters, amounting in weight to some ten tons, were conveyed through the four months, in addition to which at least an equal weight of other freight was taken up, exclusive of actual passengers, of whom no fewer than two hundred were transported from the beleaguered city. Of these only one returned, seven or eight were drowned, twice this number were taken prisoners, and as many again more or less injured in descents. From a purely financial point of view the undertaking was no failure, as the cost, great as it necessarily became, was, it is said, fairly covered by the postage, which it was possible and by no means unreasonable to levy. The recognised tariff seems to have been 20 centimes for 4 grammes, or at the rate of not greatly more than a shilling per English ounce. Surely hardly on a par with fame in prices in a time of siege.

It has already been stated that the defenders of Paris did not derive substantial assistance from the services of such a reconnoitring balloon as is generally used in warfare at every available opportunity. It is possible that the peculiar circumstances of the investment of the town rendered such reconnaissance of comparatively small value. But, at any rate, it seems clear that due opportunity was not given to this strategic method. M. Giffard, who at the commencement of the siege was in Paris, and whose experience with a captive balloon was second to none, made early overtures to the Government, offering to build for L40,000 a suitable balloon, capable of raising forty persons to a height of 3,000 feet. Forty aerial scouts, it may be said, are hardly needed for purposes of outlook at one time; but it appears that this was not the consideration which stood in the way of M. Giffard's offer being accepted. According to M. de Fonvielle, the Government refused the experienced aeronaut's proposal on the ground that he required a place in the Champs Elysees, "which it would be necessary to clear of a few shrubs"!

CHAPTER XIX. THE TRAGEDY OF THE ZENITH — THE NAVIGABLE BALLOON

The mechanical air ship had, by this time, as may be inferred, begun seriously to occupy the attention of both theoretical and practical aeronauts. One of the earliest machines deserving of special mention was designed by M. Giffard, and consisted of an elongated balloon, 104 feet in length and 39 feet in greatest diameter, furnished with a triangular rudder, and a steam engine operating a screw. The fire of the engine, which burned coke, was skilfully protected, and the fuel and water required were taken into calculation as so much ballast to be gradually expended. In this vessel, inflated only with coal gas, and somewhat unmanageable and difficult to balance, the enthusiastic inventor ascended alone from the Hippodrome and executed sundry desired movements, not unsuccessfully. But the trial was not of long duration, and the descent proved both rapid and perilous. Had the trial been made in such a perfect calm as that which prevailed when certain subsequent inventions were tested, it was considered that M. Giffard's vessel would have been as navigable as a boat in the water. This unrivalled mechanician, after having made great advances in the direction of high speed engines of sufficient lightness, proceeded to design a vastly improved dirigible balloon, when his endeavours were frustrated by blindness.

As has been already stated, M. Dupuy de Lome, at the end of the siege of Paris, was engaged in building a navigable balloon, which, owing to the unsettled state of affairs in France, did not receive its trial till two years later. This balloon, which was inflated with pure hydrogen, was of greater capacity than that of M. Giffard, being cigar shaped and measuring 118 feet by 48 feet. It was also provided with an ingenious arrangement consisting of an internal air bag, capable of being either inflated or discharged, for the purpose of keeping the principal envelope always distended, and thus offering the least possible resistance to the wind. The propelling power was the manual labour of eight men working the screw, and the steerage was provided for by a triangular rudder. The trial, which was carried out without mishap, took place in February, 1872, in the Fort of Vincennes, under the personal direction of the inventor, when it was found that the vessel readily obeyed the helm, and was capable of a speed exceeding six miles an hour.

It was not till nine years after this that the next important trial with air ships was made. The brothers Tissandier will then be found taking the lead, and an appalling incident in the aeronautical career of one of these has now to be recorded.

In the spring of 1875, and with the co-operation of French scientific societies, it was determined to make two experimental voyages in a balloon called the "Zenith," one of these to be of long duration, the other of great height. The first of these had been successfully accomplished in a flight of twenty-four hours' duration from Paris to Bordeaux. It was now April the 15th, and the lofty flight was embarked upon by M. Gaston Tissandier, accompanied by MM. Croce-Spinelli and Sivel. Under competent advice, provision for respiration on emergency was provided in three small balloons, filled with a mixture of air and oxygen, and fitted with indiarubber hose pipes, which would allow the mixture, when inhaled, to pass first through a wash bottle containing aromatic fluid. The experiments determined on included an analysis of the proportion of carbonic acid gas at different heights by means of special apparatus; spectroscopic observations, and the readings registered by certain barometers and thermometers. A novel and valuable experiment, also arranged, was that of testing the internal temperature of the balloon as compared with that of the external air.

Ascending at 11.30 a.m. under a warm sun, the balloon had by 1 p.m. reached an altitude of 16,000 feet, when the external air was at freezing point, the gas high in the balloon being 72 degrees, and at the centre 66 degrees. Ere this height had been fully reached, however, the voyagers had begun to breathe oxygen. At 11.57, an hour previously, Spinelli had written in his notebook, "Slight pain in the ears — somewhat oppressed — it is the gas." At 23,000 feet Sivel wrote in his notebook, "I am inhaling oxygen — the effect is excellent," after which he proceeded to urge the balloon higher by a discharge of ballast. The rest of the terrible narrative has now to be taken from the notes of M. Tissandier, and as these constitute one of the most thrilling narratives in aeronautical records we transcribe them nearly in full, as given by Mr. Glaisher: —

"At 23,000 feet we were standing up in the car. Sivel, who had given up for a moment, is re-invigorated. Croce-Spinelli is motionless in front of me.... I felt stupefied and frozen. I wished to put on my fur gloves, but, without being conscious of it, the action of taking them from my pocket necessitated an effort that I could no longer make.... I copy, verbatim, the following lines which were written by me, although I have no very distinct remembrance of doing so. They are traced in a hardly legible manner by a hand trembling with cold: 'My hands are frozen. I am all right. We are all all right. Fog in the horizon, with little rounded cirrus. We are ascending. Croce pants; he inhales oxygen. Sivel closes his eyes. Croce also closes his eyes.... Sivel throws out ballast'—these last words are hardly readable. Sivel seized his knife and cut successively three cords, and the three bags emptied themselves and we ascended rapidly. The last remembrance of this ascent which remains clear to me relates to a moment earlier. Croce-Spinelli was seated, holding in one hand a wash bottle of oxygen gas. His head was slightly inclined and he seemed oppressed. I had still strength to tap the aneroid barometer to facilitate the movement of the needle. Sivel had just raised his hand towards the sky. As for myself, I remained perfectly still, without suspecting that I had, perhaps, already lost the power of moving. About the height of 25,000 feet the condition of stupefaction which ensues is extraordinary. The mind and body weaken by degrees, and imperceptibly, without consciousness of it. No suffering is then experienced; on the contrary, an inner joy is felt like an irradiation from the surrounding flood of light. One becomes indifferent. One thinks no more of the perilous position or of danger. One ascends, and is happy to ascend. The vertigo of the upper regions is not an idle word; but, so far as I can judge from my personal impression, vertigo appears at the last moment; it immediately precedes annihilation, sudden, unexpected, and irresistible.

"When Sivel cut away the bags of ballast at the height of about 24,000 feet, I seemed to remember that he was sitting at the bottom of the car, and nearly in the same position as Croce-Spinelli. For my part, I was in the angle of the car, thanks to which support I was able to hold up; but I soon felt too weak even to turn my head to look at my companions. Soon I wished to take hold of the tube of oxygen, but it was impossible to raise my arm. My mind, nevertheless, was quite clear. I wished to explain, 'We are 8,000 metres high'; but my tongue was, as it were, paralysed. All at once I closed my eyes, and, sinking down inert, became insensible. This was about 1.30 p.m. At 2.8 p.m. I awoke for a moment, and found the balloon rapidly descending. I was able to cut away a bag of ballast to check the speed and write in my notebook the following lines, which I copy:

"'We are descending. Temperature, 3 degrees. I throw out ballast. Barometer, 12.4 inches. We are descending. Sivel and Croce still in a fainting state at the bottom of the car. Descending very rapidly.'

"Hardly had I written these lines when a kind of trembling seized me, and I fell back weakened again. There was a violent wind from below, upwards, denoting a very rapid descent. After some minutes I felt myself shaken by the arm, and I recognised Croce, who had revived. 'Throw out ballast,' he said to me, 'we are descending '; but I could hardly open my eyes, and did not see whether Sivel was awake. I called to mind that Croce unfastened the aspirator, which he then threw overboard, and then he threw out ballast, rugs, etc.

"All this is an extremely confused remembrance, quickly extinguished, for again I fell back inert more completely than before, and it seemed to me that I was dying. What happened? It is certain that the balloon, relieved of a great weight of ballast, at once ascended to the higher regions.

"At 3.30 p.m. I opened my eyes again. I felt dreadfully giddy and oppressed, but gradually came to myself. The balloon was descending with frightful speed and making great oscillations. I crept along on my knees, and I pulled Sivel and Croce by the arm. 'Sivel! Croce!' I exclaimed, 'Wake up!' My two companions were huddled up motionless in the car, covered by their cloaks. I collected all my strength, and endeavoured to raise them up. Sivel's face was black, his eyes dull, and his mouth was open and full of blood. Croce's eyes were half closed and his mouth was bloody.

"To relate what happened afterwards is quite impossible. I felt a frightful wind; we were still 9,700 feet high. There remained in the car two bags of ballast, which I threw out. I was drawing near the earth. I looked for my knife to cut the small rope which held the anchor, but could not find it. I was like a madman, and continued to call 'Sivel! Sivel!' By good fortune I was able to put my hand upon my knife and detach the anchor at the right moment. The shock on coming to the ground was dreadful. The balloon seemed as if it were being flattened. I thought it was going to remain where it had fallen, but the wind was high, and it was dragged across fields, the anchor not catching. The bodies of my unfortunate friends were shaken about in the car, and I thought every moment they would be jerked out. At length, however, I seized the valve line, and the gas soon escaped from the balloon, which lodged against a tree. It was then four o'clock. On stepping out, I was seized with a feverish attack, and sank down and thought for a moment that I was going to join my friends in the next world; but I came to. I found the bodies of my friends cold and stiff. I had them put under shelter in an adjacent barn. The descent of the 'Zenith' took place in the plains 155 miles from Paris as the crow flies. The greatest height attained in this ascent is estimated at 28,000 feet."

It was in 1884 that the brothers Tissandier commenced experiments with a screw-propelled air ship resembling in shape those constructed by Giffard and Dupuy de Lome, but smaller, measuring only 91 feet by 30 feet, and operated by an electric motor placed in circuit with a powerful battery of bichromate cells. Two trials were made with this vessel in October, 1883, and again in the following September, when it proved itself capable of holding its course in calm air and of being readily controlled by the rudder.

But, ere this, a number of somewhat similar experiments, on behalf of the French Government, had been entered upon by Captains Renard and Krebs at Chalais-Meudon. Their balloon may be described as fish-shaped, 165 feet long, and 27.5 feet in principal diameter. It was operated by an electric motor, which was capable of driving a screw of large dimensions at forty-eight revolutions per minute. At its first trial, in August, 1884, in dead calm, it attained a velocity of over twelve miles per hour, travelling some two and a half miles in a forward direction, when, by application of the rudder and judicious management, it was manoeuvred homewards, and practically brought to earth at the point of departure.

A more important trial was made on the 12th of the following month, and was witnessed by M. Tissandier, according to whom the aerostat conveying the inventors ascended gently and steadily, drifting with an appreciable breeze until the screw was set in motion and the helm put down, when the vessel was brought round to the wind and held its own until the motor, by an accident, ceased working. A little later the same air ship met with more signal success. On one occasion, starting from Chalais-Meudon, it took a direct course to the N.E., crossing the railway and the Seine, where the aeronauts, stopping the screw, ascertained the velocity of the wind to be approximately five miles an hour. The screw being again put in motion, the balloon was steered to the right, and, following a path parallel to its first, returned to its point of departure. Starting again the same afternoon, it was caused to perform a variety of aerial evolutions, and after thirty-five minutes returned once more to its starting place.

A tabular comparison of the four navigable balloons which we have now described has been given as follows:—

Date. Name. Motor. Vel. p. Sec. 1852 M.
Henri Giffard Steam engine 13.12 ft. 1872 M.
Dupuy de Lome Muscular force 9.18 ft. 1883
MM. Tissandier Electric motor 9.84 ft. 1884
MM. Renard & Krebs Electric motor 18.04 ft.

About this period, that is in 1883, and really prior to the Meudon experiments, there were other attempts at aerial locomotion not to be altogether passed over, which were made also in France, but financed by English money. The experiments were performed by Mr. F. A. Gower, who, writing to Professor Tyndall, claims to have succeeded in "driving a large balloon fairly against the wind by steam power." A melancholy interest will always belong to these trials from the fact that Mr. Gower was subsequently blown out to sea with his balloon, leaving no trace behind.

At this stage it will be well to glance at some of the more important theories which were being mooted as to the possibility of aerial locomotion properly so called. Broadly, there were two rival schools at this time. We will call them the "lighter-than-air-ites" and the "heavier-than-air-ites," respectively. The former were the advocates of the air vessel of which the balloon is a type. The latter school maintained that, as birds are heavier than air, so the air locomotive of the future would be a machine itself heavier than air, but capable of being navigated by a motor yet to be discovered, which would develop proportionate power. Sir H. Maxim's words may be aptly quoted here. "In all Nature," he says, "we do not find a single balloon. All Nature's flying machines are heavier than the air, and depend altogether upon the development of dynamic energy."

The faculty of soaring, possessed by many birds, of which the albatross may be considered a type, led to numerous speculations as to what would constitute the ideal principle of the air motor. Sir G. Cayley, as far back as 1809, wrote a classical article on this subject, without, however, adding much to its elucidation. Others after his time conceived that the bird, by sheer habit and practice, could perform, as it were, a trick in balancing by making use of the complex air streams varying in speed and direction that were supposed to intermingle above.

Mr. R. A. Proctor discusses the matter with his usual clear-sightedness. He premises that the bird may, in actual fact, only poise itself for some ten minutes—an interval which many will consider far too small—without flap of the wings, and, while contending that the problem must be simply a mechanical one, is ready to admit that "the sustaining power of the air on bodies of a particular form travelling swiftly through it may be much greater or very different in character from what is supposed." In his opinion, it is a fact that a flat body travelling swiftly and horizontally will sink towards the ground much more slowly than a similar body moving similarly but with less speed. In proof of this he gives the homely illustration of a flat stone caused to make "ducks and drakes." Thus he contends that the bird accomplishes its floating feat simply by occasional powerful propulsive efforts, combined with perfect balance. From which he deduces the corollary that "if ever the art of flying, or rather of making flying machines, is attained by man, it will be by combining rapid motion with the power of perfect balancing."

It will now appear as a natural and certain consequence that a feature to be introduced by experimentalists into flying machines should be the "Aeroplane," or, in other words, a plane which, at a desired angle, should be driven at speed through the air. Most notable attempts with this expedient were now shortly made by Hiram Maxim, Langley, and others.

But, contemporaneously with these attempts, certain feats with the rival aerostat—the balloon—were accomplished, which will be most fittingly told in this place.

CHAPTER XX. A CHAPTER OF ACCIDENTS.

It will have been gathered from what has been already stated that the balloonist is commonly in much uncertainty as to his precise course when he is above the clouds, or when unable from darkness to see the earth beneath him. With a view of overcoming this disadvantage some original experiments were suggested by a distinguished officer, who during the seventies had begun to interest himself in aeronautics.

This was Captain Burnaby. His method was to employ two small silk parachutes, which, if required, might carry burning magnesium wires, and which were to be attached to each other by a length of silk thread. On dropping one parachute, it would first partake of the motion of the balloon, but would presently drop below, when the second parachute would be dismissed, and then an imaginary line drawn between the two bodies was supposed to betray the balloon's course. It should be mentioned, however, that if a careful study is made of the course of many descending parachutes it will be found that their behaviour is too uncertain to be relied upon for such a purpose as the above. They will often float behind the balloon's wake, but sometimes again will be found in front, and sometimes striking off in some side direction, so wayward and complex are the currents which control such small bodies. Mr. Glaisher has stated that a balloon's course above the clouds may be detected by observing the grapnel, supposed to be hanging below the car, as this would be seen to be out of the vertical as the balloon drifted, and thus serve to indicate the course. However this may be, the most experienced sky sailors will be found to be in perplexity as to their direction, as also their speed, when view of the earth is obscured.

But Captain Burnaby is associated notably with the adventurous side of ballooning, the most famous of his aerial exploits being, perhaps, that of crossing the English Channel alone from Dover on March 23rd, 1882. Outwardly, he made presence of sailing to Paris by sky to dine there that evening; inwardly, he had determined to start simply with a wind which bid fair for a cross-Channel trip, and to take whatever chances it might bring him.

Thus, at 10.30 a.m., just as the mail packet left the pier, he cast off with a lifting power which rapidly carried him to a height of 2,000 feet, when he found his course to be towards Folkestone. But by shortly after 11 o'clock he had decided that he was changing his direction, and when, as he judged, some seven miles from Boulogne, the wind was carrying him not across, but down the Channel. Then, for nearly four hours, the balloon shifted about with no improvement in the outlook, after which the wind fell calm, and the balloon remained motionless at 2,000 feet above the sea. This state of things continuing for an hour, the Captain resolved on the heroic expedient of casting out all his ballast and philosophically abiding the issue. The manoeuvre turned out a happy one, for the balloon, shooting up to 11,000 feet, caught a current, on which it was rapidly carried towards and over the main land; and, when twelve miles beyond Dieppe, it became easy to descend to a lower level by manipulation of the valve, and finally to make a successful landing in open country beyond.

A few years before, an attempt to cross the Channel from the other side ended far more disastrously. Jules Duruof, already mentioned as having piloted the first runaway balloon from beleaguered Paris, had determined on an attempt to cross over to England from Calais; and, duly advertising the event, a large concourse assembled on the day announced, clamouring loudly for the ascent. But the wind proved unsuitable, setting out over the North Sea, and the mayor thought fit to interfere, and had the car removed so as to prevent proceedings. On this the crowd grew impatient, and Duruof, determining to keep faith with them, succeeded by an artifice in regaining his car, which he hastily carried back to the balloon, and immediately taking his seat, and accompanied by his wife, the intrepid pair commenced their bold flight just as the shades of evening were settling down. Shortly the balloon disappeared into the gathering darkness, and then for three days Calais knew no more of balloon or balloonists.

Neither could the voyagers see aught for certain of their own course, and thus through the long night hours their attention was wholly needed, without chance of sleep, in closely watching their situation, lest unawares they should be borne down on the waves. When morning broke they discovered that they were still being carried out over the sea on a furious gale, being apparently off the Danish coast, with the distant mountains of Norway dimly visible on the starboard bow. It was at this point, and possibly owing to the chill commonly experienced aloft soon after dawn, that the balloon suddenly took a downward course and plunged into the sea, happily, however, fairly in the track of vessels. Presently a ship came in sight, but cruelly kept on its course, leaving the castaways in despair, with their car fast succumbing to the waves.

Help, nevertheless, was really at hand. The captain of an English fishing smack, the Grand Charge, had sighted the sinking balloon, and was already bearing down to the rescue. It is said that when, at length, a boat came alongside as near as it was possible, Madame Duruof was unable to make the necessary effort to jump on board, and her husband had to throw her into the arms of the sailors. A fitting sequel to the story comes from Paris, where the heroic couple, after a sojourn in England, were given a splendid reception and a purse of money, with which M. Duruof forthwith constructed a new balloon, named the "Ville de Calais."

On the 4th of March, 1882, the ardent amateur balloonist, Mr. Simmons, had a narrow escape in circumstances somewhat similar to the above. He was attempting, in company with Colonel Brine, to cross the Channel from Canterbury, when a change of wind carried them out towards the North Sea. Falling in the water, they abandoned their balloon, but were rescued by the mail packet Foam.

The same amateur aeronaut met with an exciting experience not long after, when in company with Sir Claude C. de Crespigny. The two adventurers left Maldon, in Essex, at 11 p.m., on an August night, and, sailing at a great height out to sea, lost all sight of land till 6 a.m. the next morning, when, at 17,000 feet altitude, they sighted the opposite coast and descended in safety near Flushing.

Yet another adventure at sea, and one which proved fatal and unspeakably regrettable, occurred about this time, namely, on the 10th of December, 1881, when Captain Templer, Mr. W. Powell, M.P., and Mr. Agg-Gardner ascended from Bath. We prefer to give the account as it appears in a leading article in the Times for December 13th of that year.

After sailing over Glastonbury, "Crewkerne was presently sighted, then Beaminster. The roar of the sea gave the next indication of the locality to which the balloon had drifted and the first hint of the possible perils of the voyage. A descent was now effected to within a few hundred feet of earth, and an endeavour was made to ascertain the exact position they had reached. The course taken by the balloon between Beaminster and the sea is not stated in Captain Templer's letter. The wind, as far as we can gather, must have shifted, or different currents of air must have been found at the different altitudes. What Captain Templer says is that they coasted along to Symonsbury, passing, it would seem, in an easterly direction and keeping still very near to the earth. Soon after they had left Symonsbury, Captain Templer shouted to a man below to tell them how far they were from Bridport, and he received for answer that Bridport was about a mile off. The pace at which the balloon was moving had now increased to thirty-five miles an hour. The sea was dangerously close, and a few minutes in a southerly current of air would have been enough to carry them over it. They seem, however, to have been confident in their own powers of management. They threw out ballast, and rose to a height of 1,500 feet, and thence came down again only just in time, touching the ground at a distance of about 150 yards from the cliff. The balloon here dragged for a few feet, and Captain Templer, who had been letting off the gas, rolled out of the car, still holding the valve line in his hand. This was the last chance of a safe escape for anybody. The balloon, with its weight lightened, went up about eight feet. Mr. Agg-Gardner dropped out and broke his leg. Mr. Powell now remained as the sole occupant of the car. Captain Templer, who had still hold of the rope, shouted to Mr. Powell to come down the line. This he attempted to do, but in a few seconds, and before he could commence his perilous descent, the line was torn out of Captain Templer's hands. All communication with the earth was cut off, and the balloon rose rapidly, taking Mr. Powell with it in a south-easterly direction out to sea."

It was a few seasons previous to this, namely, on the 8th of July, 1874, when Mr. Simmons was concerned in a balloon fatality of a peculiarly distressing nature. A Belgian, Vincent de Groof, styling himself the "Flying Man," announced his intention of descending in a parachute from a balloon piloted by Mr. Simmons, who was to start from Cremorne Gardens. The balloon duly ascended, with De Groof in his machine suspended below, and when over St. Luke's Church, and at a height estimated at 80 feet, it is thought that the unfortunate man overbalanced himself after detaching his apparatus, and fell forward, clinging to the ropes. The machine failed to open, and De Groof was precipitated into Robert Street, Chelsea, expiring almost immediately. The porter of Chelsea Infirmary, who was watching the balloon, asserted that he fancied the falling man called out twice, "Drop into the churchyard; look out!" Mr. Simmons, shooting upwards in his balloon, thus suddenly lightened, to a great height, became insensible, and when he recovered consciousness found himself over Victoria Park. He made a descent, without mishap, on a line of railway in Essex.

On the 19th of August, 1887, occurred an important total eclipse of the sun, the track of which lay across Germany, Russia, Western Siberia, and Japan. At all suitable stations along the shadow track astronomers from all parts of the world established themselves; but at many eclipses observers had had bad fortune owing to the phenomenon at the critical moment being obscured. And on this account one astronomer determined on measures which should render his chances of a clear view a practical certainty. Professor Mendeleef, in Russia, resolved to engage a balloon, and by rising above the cloud barrier, should there be one, to have the eclipse all to himself. It was an example of fine enthusiasm, which, moreover, was presently put to a severe and unexpected test, for the balloon, when inflated, proved unable to take up both the aeronaut and the astronomer, whereupon the latter, though wholly inexperienced, had no alternative but to ascend alone, which, either by accident or choice, he actually did. Shooting up into space, he soon reached an altitude of 11,500 feet, where he obtained, even if he did not enjoy, an unobstructed view of the Corona. It may be supposed, however, that, owing to the novelty of his situation, his scientific observations may not have been so complete as they would have been on terra firma.

In the same month an attempt to reach a record height was made by MM. Jovis and Mallet at Paris, with the net result that an elevation of 23,000 feet was reached. It will have been noted that the difficulty through physical exhaustion of inhaling oxygen from either a bag or cylinder is a serious matter not easily overcome, and it has been suggested that the helmet invented by M. Fleuss might prove of value. This contrivance, which has scarcely attracted the attention it has merited, provides a receptacle for respiration, containing oxygen and certain purifying media, by means of which the inventor was able to remain for hours under water without any communication with the outward air.

About the period at which we have now arrived two fatal accidents befel English aeronauts. We have related how Maldon, in Essex, was associated with one of the more adventurous exploits in Mr. Simmons's career. It was fated also to be associated with the voyage with which his career closed. On August 27th, 1888, he ascended from Olympia in company with Mr. Field, of West Brighton, and Mr. Myers, of the Natural History Museum, with the intention, if practicable, of crossing to Flanders; and the voyage proceeded happily until the neighbourhood of Maldon was reached, when, as the sea coast was in sight, and it was already past five o'clock, it appeared prudent to Mr. Simmons to descend and moor the balloon for the night. Some labourers some three miles from Maldon sighted the balloon coming up at speed, and at the same time descending until its grapnel commenced tearing through a field of barley, when ballast was thrown out, causing the balloon to rise again towards and over some tall elms, which became the cause of the disaster which followed. The grapnel, catching in the upper boughs of one of these trees, held fast, while the balloon, borne by the force of a strong wind, was repeatedly blown down to earth with violence, rebounding each time to a considerable height, only to be flung down again on the same spot. After three or four impacts the balloon is reported to have burst with a loud noise, when high in the air, the silk being blown about over the field, and the car and its occupants dashed to the ground. Help was unavailing till this final catastrophe, and when, at length, the labourers were able to extricate the party, Mr. Simmons was found with a fractured skull and both companions badly injured.

Four summers later, June 30th, 1892, Captain Dale, the aeronaut to the Crystal Palace, was announced to make an ascent from the usual balloon grounds, weather permitting. Through the night and morning a violent storm prevailed, and it was contemplated that the exhibition would be withdrawn; but the wind abating in the afternoon, the inflation was proceeded with, and the ascent took place shortly before 6 p.m., not, however, before a large rent had been discovered and repaired as far as possible by Mrs. Dale. As passengers, there ascended the Captain's son William, aged nineteen, Mr. J. Macintosh, and Mr. Cecil Shadbolt. When the balloon had reached an altitude estimated at 600 feet the onlookers were horrified to see it suddenly collapse, a large rent having developed near the top part of the silk, from which the gas "rushed out in a dense mass, allowing the balloon to fall like a rag." The occupants of the car were seen to be throwing out everything madly, even wrenching the buttons from their clothing. All, however, with little avail, for the balloon fell "with a sickening thud," midway between the Maze and lower lake. All were found alive; but Captain Dale, who had alighted on his back, died in a few minutes; Mr. Shadbolt succumbed later, and both remaining passengers sustained terrible injuries.

Few balloon mishaps, unattended with fatal results, have proved more exciting than the following. A large party had ascended from Belfast, in a monster balloon, under the guidance of Mr.

Coxwell, on a day which was very unfit for the purpose by reason of stormy weather. A more serious trouble than the wind, however, lay in several of the passengers themselves, who seem to have been highly excitable Irishmen, incapable at the critical moment of quietly obeying orders.

The principal hero of the story, a German. Mr. Runge, in writing afterwards to the Ulster Observer, entirely exonerates Mr. Coxwell from any blame, attributing his mischances solely to the reprehensible conduct of his companions. On approaching the ground, Mr. Coxwell gave clear instructions. The passengers were to sit down in an unconstrained position facing each other, and be prepared for some heavy shocks. Above all things they were to be careful to get out one by one, and on no account to leave hold of the car. Many of the passengers, however, refused to sit down, and, according to Mr. Runge, "behaved in the wildest manner, losing completely their self-control. Seizing the valve rope themselves, they tore it away from its attachment, the stronger pushing back the weaker, and refusing to lend help when they had got out. In consequence of this the car, relieved of their weight, tore away from the grasp of Mr. Coxwell and those who still clung to it, and rose above the trees, with Mr. Runge and one other passenger, Mr. Halferty, alone within. As the balloon came earthwards again, they shouted to the countrymen for succour, but without the slightest avail, and presently, the anchor catching, the car struck the earth with a shock which threw Mr. Halferty out on the ground, leaving Mr. Runge to rise again into the air, this time alone." He thus continues the story:—

"The balloon moved on, very soon, in a horizontal direction straight towards the sea, which we were then rapidly nearing. Coming to a farm, I shouted out to the people standing there. Some women, with their quick humane instincts, were the first to perceive my danger, and exhorted the men to hurry to my assistance, they themselves running as fast as they could to tender what little help they might be able to give me. The anchor stuck in a willow tree. I shouted out to the people below to secure the cable and anchor by ropes, which they did. The evening was now beautifully still, the breeze had died away, and the balloon was swinging calmly at her moorings above the farmhouse. One of the men asked me whether I had a rope with me, and how I intended to get out. I told them only to take care of the cable, because the balloon would settle down by herself before long. I was congratulating myself on a speedy escape from my dangerous position. I had not counted on the wind. A breeze in about six or eight minutes sprang up, tossed the balloon about like a large sail, then a crash, and — the anchor was loose again. It tore through the trees, flinging limbs and branches about like matches. It struck the roof of the farmhouse, splintering the chimneys and tiles like glass.

"On I went; I came near another farm; shouted out for help, and told the men to secure the anchor to the foot of a large tree close by. The anchor was soon made fast, but this was only a momentary relief. The breeze again filled the half-empty balloon like a sail, there was a severe strain on the cable, then a dull sound, and a severe concussion of the basket — the cable, strange fatality, had broken, and the anchor, my last and only hope, was gone. I was now carried on in a straight direction towards the sea, which was but a short distance ahead. The anchor being lost I gave up all hope. I sat down resigned in the car, and prepared for the end. All at once I discovered that a side current was drifting me towards the mountain; the car struck the ground, and was dashing along at a fearful rate, knocking down stone fences and breaking everything it came in contact with in its wild career. By-and-by the knocks became less frequent. We were passing over a cultivated country, and the car was, as it were skimming the surface and grazing the top of the hedges. I saw a thick hawthorn hedge at some distance before me, and the balloon rapidly sweeping towards it. That was my only chance. I rushed to the edge of the car and flung myself down upon the hedge."

CHAPTER XXI. THE COMING OF THE FLYING MACHINE.

In the early nineties the air ship was engaging the attention of many inventors, and was making important strides in the hands of Mr. Maxim. This unrivalled mechanician, in stating the case, premises that a motive power has to be discovered which can develop at least as much power in proportion to its weight as a bird is able to develop. He asserts that a heavy bird, with relatively small wings—such as a goose—carries about 150 lb. to the horse power, while the albatross or the vulture, possessed of proportionately greater winged surface, can carry about 250 lbs. per horse power.

Professor Langley, of Washington, working contemporaneously, but independently of Mr. Maxim, had tried exhaustive experiments on a rotating arm (characteristically designated by Mr. Maxim a "merry-go-round"), thirty feet long, applying screw propellers. He used, for the most part, small planes, carrying loads of only two or three pounds, and, under these circumstances, the weight carried was at the rate of 250 lbs. per horse power. His important statements with regard to these trials are that one-horse power will transport a larger weight at twenty miles an hour than at ten, and a still larger at forty miles than at twenty, and so on; that "the sustaining pressure of the air on a plane moving at a small angle of inclination to a horizontal path is many times greater than would result from the formula implicitly given by Newton, while, whereas in land or marine transport increased speed is maintained only by a disproportionate expenditure of power within the limits of experiment, in aerial horizontal transport the higher speeds are more economical of power than the lower ones."

This Mr. Maxim is evidently ready to endorse, stating, in his own words, that birds obtain the greater part of their support by moving forward with sufficient velocity so as to be constantly resting on new air, the inertia of which has not been disturbed. Mr. Maxim's trials were on a scale comparable with all his mechanical achievements. He employed for his experiments a rotating arm, sweeping out a circle, the circumference of which was 200 feet. To the end of this arm he attached a cigar-shaped apparatus, driven by a screw, and arranged in such a manner that aeroplanes could be attached to it at any angle. These planes were on a large scale, carrying weights of from 20 lbs. to 100 lbs. With this contrivance he found that, whatever push the screw communicated to the aero-plane, "the plane would lift in a vertical direction from ten to fifteen times as much as the horizontal push that it received from the screw, and which depended upon the angle at which the plane was set, and the speed at which the apparatus was travelling through the air." Next, having determined by experiment the power required to perform artificial flight, Mr. Maxim applied himself to designing the requisite motor. "I constructed," he states, "two sets of compound engines of tempered steel, all the parts being made very light and strong, and a steam generator of peculiar construction, the greater part of the heating surface consisting of small and thin copper tubes. For fuel I employed naphtha."

This Mr. Maxim wrote in 1892, adding that he was then experimenting with a large machine, having a spread of over 100 feet. Labour, skill, and money were lavishly devoted henceforward to the great task undertaken, and it was not long before the giant flying machine, the outcome of so much patient experimenting, was completed and put to a practical trial. Its weight was 7,500 lbs. The screw propellers were nearly 18 feet in diameter, each with two blades, while the engines were capable of being run up to 360 horse power. The entire machine was mounted on an inner railway track of 9 feet and an outer of 35 feet gauge, while above there was a reversed rail along which the machine would begin to run so soon as with increase of speed it commenced to lift itself off the inner track.

In one of the latest experiments it was found that when a speed of 42 miles an hour was attained all the wheels were running on the upper track, and revolving in the opposite direction from those on the lower track. However, after running about 1,000 feet, an axle tree doubled up, and immediately afterwards the upper track broke away, and the machine, becoming liberated, floated in the air, "giving those on board a sensation of being in a boat."

The experiment proved conclusively to the inventor that a machine could be made on a large scale, in which the lifting effect should be considerably greater than the weight of the machine, and this, too, when a steam engine was the motor. When, therefore, in the years shortly following, the steam engine was for the purposes of aerial locomotion superseded by the lighter and more suitable petrol engine, the construction of a navigable air ship became vastly more practicable. Still, in Sir H. Maxim's opinion, lately expressed, "those who seek to navigate the air by machines lighter than the air have come, practically, to the end of their tether," while, on the other hand, "those who seek to navigate the air with machines heavier than the air have not even made a start as yet, and the possibilities before them are very great indeed."

As to the assertion that the aerial navigators last mentioned "have not even made a start as yet," we can only say that Sir H. Maxim speaks with far too much modesty. His own colossal labours in the direction of that mode of aerial flight, which he considers to be alone feasible, are of the first importance and value, and, as far as they have gone, exhaustive. Had his experiments been simply confined to his classical investigations of the proper form of the screw propeller his name would still have been handed down as a true pioneer in aeronautics. His work, however, covers far wider ground, and he has, in a variety of ways, furnished practical and reliable data, which must always be an indispensable guide to every future worker in the same field.

Professor Langley, in attacking the same problem, first studied the principle and behaviour of a well-known toy — the model invented by Penaud, which, driven by the tension of india-rubber, sustains itself in the air for a few seconds. He constructed over thirty modifications of this model, and spent many months in trying from these to as certain what he terms the "laws of balancing leading to horizontal flight." His best endeavours at first, however, showed that he needed three or four feet of sustaining surface to a pound of weight, whereas he calculated that a bird could soar with a surface of less than half a foot to the pound. He next proceeded to steam-driven models in which for a time he found an insuperable difficulty in keeping down the weight, which, in practice, always exceeded his calculation; and it was not till the end of 1893 that he felt himself prepared for a fair trial. At this time he had prepared a model weighing between nine and ten pounds, and he needed only a suitable launching apparatus to be used over water. The model would, like a bird, require an initial velocity imparted to it, and the discovery of a suitable apparatus gave him great trouble. For the rest the facilities for launching were supplied by a houseboat moored on the Potomac. Foiled again and again by many difficulties, it was not till after repeated failures and the lapse of many months, when, as the Professor himself puts it, hope was low, that success finally came. It was in the early part of 1896 that a successful flight was accomplished in the presence of Dr. Bell, of telephone fame, and the following is a brief epitome of the account that this accomplished scientist contributed to the columns of Nature:—

"The flying machine, built, apparently, almost entirely of metal, was driven by an engine said to weigh, with fuel and water, about 25 lbs., the supporting surface from tip to tip being 12 or 14 feet. Starting from a platform about 20 feet high, the machine rose at first directly in the face of the wind, moving with great steadiness, and subsequently wheeling in large curves until steam was exhausted, when, from a height of 80 or 100 feet, it shortly settled down. The experiment was then repeated with similar results. Its motion was so steady that a glass of water might have remained unspilled. The actual length of flight each time, which lasted for a minute and a half, exceeded half a mile, while the velocity was between twenty and twenty-five miles an hour in a course that was constantly taking it 'up hill.' A yet more successful flight was subsequently made."

But flight of another nature was being courageously attempted at this time. Otto Lilienthal, of Berlin, in imitation of the motion of birds, constructed a flying apparatus which he operated himself, and with which he could float down from considerable elevations. "The feat," he warns tyros, "requires practice. In the beginning the height should be moderate, and the wings not too large, or the wind will soon show that it is not to be trifled with." The inventor commenced with all due caution, making his first attempt over a grass plot from a spring board one metre high, and subsequently increasing this height to two and a half metres, from which elevation he could safely cross the entire grass plot. Later he launched himself from the lower ridges of a hill 250 feet high, when he sailed to a distance of over 250 yards, and this time he writes enthusiastically of his self-taught accomplishment: —

"To those who, from a modest beginning and with gradually increased extent and elevation of flight have gained full control over the apparatus, it is not in the least dangerous to cross deep and broad ravines. It is a difficult task to convey to one who has never enjoyed aerial flight a clear perception of the exhilarating pleasure of this elastic motion. The elevation above the ground loses its terrors, because we have learned by experience what sure dependence may be placed upon the buoyancy of the air."

As a commentary to the above we extract the following: — "We have to record the death of Otto Lilienthal, whose soaring machine, during a gliding flight, suddenly tilted over at a height of about 60 feet, by which mishap he met an untimely death on August 9th, 1896." Mr. O. Chanute, C.E. of Chicago, took up the study of gliding flight at the point where Lilienthal left it, and, later, Professor Fitzgerald and others. Besides that invented by Penaud, other aero-plane models demanding mention had been produced by Tatin, Moy, Stringfellow, and Lawrence Hargrave, of Australia, the subsequent inventor of the well-known cellular kite. These models, for the most part, aim at the mechanical solution of the problem connected with the soaring flight of a bird.

The theoretical solution of the same problem had been attacked by Professor Langley in a masterly monograph, entitled "The Internal Work of the Wind." By painstaking experiment with delicate instruments, specially constructed, the Professor shows that wind in general, so far from being, as was commonly assumed, mere air put in motion with an approximately uniform velocity in the same strata, is, in reality, variable and irregular in its movements beyond anything which had been anticipated, being made up, in fact, of a succession of brief pulsations in different directions, and of great complexity. These pulsations, he argues, if of sufficient amplitude and frequency, would be capable, by reason of their own "internal work," of sustaining or even raising a suitably curved surface which was being carried along by the main mean air stream. This would account for the phenomenon of "soaring." Lord Rayleigh, discussing the same problem, premises that when a bird is soaring the air cannot be moving uniformly and horizontally. Then comes the natural question, Is it moving in ascending currents? Lord Rayleigh has frequently noticed such currents, particularly above a cliff facing the wind. Again, to quote another eminent authority, Major Baden-Powell, on an occasion when flying one of his own kites, found it getting to so high an angle that it presently rose absolutely overhead, with the string perpendicular. He then took up a heavy piece of wood, which, when tied to the string, began to rise in the air. He satisfied himself that this curious result was solely due to a strong uptake of the air.

But, again, Lord Rayleigh, lending support to Professor Langley's argument, points out that the apparent cause of soaring may be the non-uniformity of the wind. The upper currents are generally stronger than the lower, and it is mechanically possible for a bird, taking advantage of two adjacent air streams, different in velocity, to maintain itself in air without effort on its own part.

Lord Rayleigh, proceeding to give his views on artificial flight, declares the main problem of the flying machine to be the problem of the aerial plane. He states the case thus: — "Supposing a plane surface to be falling vertically at the rate of four miles an hour, and also moving horizontally at the rate of twenty miles an hour, it might have been supposed that the horizontal motion would make no difference to the pressure on its under surface which the falling plane must experience. We are told, however, that in actual trial the horizontal motion much increases the pressure under the falling plane, and it is this fact on which the possibility of natural and artificial flight depends."

Ere this opinion had been stated by Lord Rayleigh in his discourse on "Flight," at the Royal Institution, there were already at work upon the aero-plane a small army of inventors, of whom it will be only possible in a future chapter to mention some. Due reference, however, should here be made to Mr. W. F. Wenham, of Boston, U.S.A., who had been at work on artificial flight for many years, and to whose labours in determining whether man's power is sufficient to raise his own weight Lord Rayleigh paid a high tribute. As far back as 1866 Mr. Wenham had published a paper on aerial locomotion, in which he shows that any imitation by man of the far-extended wings of a bird might be impracticable, the alternative being to arrange the necessary length of wing as a series of aero-planes, a conception far in advance of many theorists of his time.

But there had been developments in aerostation in other lines, and it is time to turn from the somewhat tedious technicalities of mechanical flight and the theory or practice of soaring, to another important means for traversing the air—the parachute. This aerial machine, long laid aside, was to lend its aid to the navigation of the air with a reliability never before realised. Professor Baldwin, as he was termed, an American aeronaut, arrived in England in the summer of 1888, and commenced giving a series of exhibitions from the Alexandra Palace with a parachute of his own invention, which, in actual performance, seems to have been the most perfect instrument of the kind up to that time devised. It was said to be about 18 feet in diameter, whereas that of Garnerin, already mentioned, had a diameter of some 30 feet, and was distinctly top-heavy, owing to its being thus inadequately ballasted; for it was calculated that its enormous size would have served for the safe descent, not of one man, but of four or five. Baldwin's parachute, on the contrary, was reckoned to give safe descent to 250 lbs., which would include weight of man and apparatus, and reduce the ultimate fall to one not exceeding 8 feet. The parachute was attached to the ring of a small balloon of 12,000 cubic feet, and the Professor ascended, sitting on a mere sling of rope, which did duty for a car.

Mr. Thomas Moy, who investigated the mechanics of the contrivance, estimated that after a drop of 16 feet, the upward pressure, amounting to over 2 lb. per square foot, would act on a surface of not less than 254 square feet. There was, at the time, much foolish comment on the great distance which the parachute fell before it opened, a complete delusion due to the fact that observers failed to see that at the moment of separation the balloon itself sprang upward.

CHAPTER XXII. THE STORY OF THE SPENCERS.

It has been in the hands of the Spencers that the parachute, as also many other practical details of aeronautics, has been perfected, and some due sketch of the career of this family of eminent aeronauts must be no longer delayed.

Charles Green had stood godfather to the youngest son of his friend and colleague, Mr. Edward Spencer, and in later years, as though to vindicate the fact, this same son took up the science of aeronautics at the point where his father had left it. We find his name in the records of the Patent Office of 1868 as the inventor of a manumetric flying machine, and there are accounts of the flying leaps of several hundred feet which he was enabled to take by means of the machine he constructed. Again, in 1882 we find him an inventor, this time of the patent asbestos fire balloon, by means of which the principal danger to such balloons was overcome.

At this point it is needful to make mention of the third generation—the several sons who early showed their zeal and aptitude for perpetuating the family tradition. It was from his school playground that the eldest son, Percival, witnessed with intense interest what appeared like a drop floating in the sky at an immense altitude. This proved to be Simmons's balloon, which had just risen to a vast elevation over Cremorne Gardens, after having liberated the unfortunate De Groof, as mentioned in a former chapter. And one may be sure that the terrible reality of the disaster that had happened was not lost on the young school-boy. But his wish was to become an aeronauts, and from this desire nothing deterred him, so that school days were scarcely over before he began to accompany his father aloft, and in a very few years, i.e. in 1888, he had assumed the full responsibilities of a professional balloonist.

It was in this year that Professor Baldwin appeared in England, and it is easy to understand that the parachute became an object of interest to the young Spencer, who commenced on his own account a series of trials at the Alexandra Palace, and it was now, also, that chance good fortune came his way. An Indian gentleman, who was witness of his experiments, and convinced that a favourable field for their further development existed in his own country, proposed to the young aspirant that he should accompany him to India, with equipment suited for the making of a successful campaign.

Thus it came about that in the early days of 1889, in the height of the season, Mr. Percival Spencer arrived at Bombay, and at once commenced professional business in earnest. Coal gas being here available, a maiden ascent was quickly arranged, and duly announced to take place at the Government House, Paral, the chief attraction being the parachute descent, the first ever attempted in India.

This preliminary exhibition proving in all ways a complete success, Mr. Spencer, after a few repetitions of his performance, repaired to Calcutta; but here great difficulties were experienced in the matter of gas. The coal gas available was inadequate, and when recourse was had to pure hydrogen the supply proved too sluggish. At the advertised hour of departure the balloon was not sufficiently inflated, while the spectators were growing impatient. It was at this critical moment that Mr. Spencer resolved on a surprise. Suddenly casting off the parachute, and seated on a mere sling below the half-inflated balloon, without ballast, without grapnel, and unprovided with a valve, he sailed away over the heads of the multitude.

The afternoon was already far advanced, and the short tropical twilight soon gave way to darkness, when the intrepid voyager disappeared completely from sight. Excitement was intense that night in Calcutta, and greater still the next day when, as hour after hour went by, no news save a series of wild and false reports reached the city. Trains arriving from the country brought no intelligence, and telegraphic enquiries sent in all directions proved fruitless. The Great Eastern Hotel, where the young man had been staying, was literally besieged for hours by a large crowd eager for any tidings. Then the Press gave expression to the gloomiest forebodings, and the town was in a fever of unrest. From the direction the balloon had taken it was thought that, even if the aeronaut had descended in safety, he could only have been landed in the jungle of the Sunderbunds, beset with perils, and without a chance of succour. A large reward was offered for reliable information, and orders were issued to every likely station to organise a search. But ere this was fully carried into effect messages were telegraphed to England definitely asserting that Mr. Spencer had lost his life. For all this, after three days he returned to Calcutta, none the worse for the exploit.

Then the true tale was unravelled. The balloon had changed its course from S.E. to E. after passing out of sight of Calcutta, and eventually came to earth the same evening in the neighbourhood of Hossainabad, thirty-six miles distant. During his aerial flight the voyager's main trouble had been caused by his cramped position, the galling of his sling seat, and the numbing effect of cold as he reached high altitudes; but, as twilight darkened into gloom, his real anxiety was with respect to his place of landing, for he could with difficulty see the earth underneath. He heard the distant roll of the waters, caused by the numerous creeks which intersect the delta of the Ganges, and when darkness completely shut out the view it was impossible to tell whether he was over land or sea. Fortune favoured him, however, and reaching dry ground, he sprang from his seat, relinquishing at the same moment his hold of the balloon, which instantly disappeared into the darkness.

Then his wanderings began. He was in an unknown country, without knowledge of the language, and with only a few rupees in his pocket. Presently, however, seeing a light, he proceeded towards it, but only to find himself stopped by a creek. Foiled more than once in this way, he at length arrived at the dwelling of a family of natives, who promptly fled in terror. To inspire confidence and prove that he was mortal, Mr. Spencer threw his coat over the mud wall of the compound, with the result that, after examination of the garment, he was received and cared for in true native fashion, fed with rice and goat's milk, and allowed the use of the verandah to sleep in.

He succeeded in communing with the natives by dint of lead pencil sketches and dumb show, and learned, among other things, that he had descended in a little clearing surrounded by woods, and bounded by tidal creeks, which were infested with alligators. Yet, in the end, the waterways befriended him; for, as he was being ferried across, he chanced on his balloon sailing down on the tide, recovered it, and used the tidal waters for the return journey.

The greeting upon his arrival in Calcutta was enthusiastic beyond description from both Europeans and natives. The hero of the adventure was visited by rajahs and notables, who vied with each other in expressions of welcome, in making presents, even inviting him to visit the sacred precincts of their zenanas. The promised parachute descent was subsequently successfully made at Cossipore, and then followed a busy, brilliant season, after which the wanderer returned to England. By September he is in Dublin, and makes the first parachute descent ever witnessed in Ireland; but by November he is in Bombay again, whence, proceeding to Calcutta, he repeats his success of the year before. Next he visits Allahabad, where the same fortune attends him, though his balloon flies away in a temporary escape into the Jumna. By May he is ascending at

Singapore, armed here, however, with a cork jacket.

Hence, flushed with success, he repairs to the Dutch Indies, and demonstrates to the Dutch officers the use of the balloon in war. As a natural consequence, he is moved up to the seat of the Achinese War in Sumatra, where, his balloon being moored to the rear of an armoured train, an immediate move is made to the front, and orders are forthwith telephoned from various centres to open fire on the enemy. Mr. Spencer, the while accompanied by an officer, makes a captive ascent, in which for some time he is actually under the enemy's fire. The result of this plucky experiment is a most flattering official report. In all the above-mentioned ascents he made his own gas without a hitch.

Thence he travels on with the same trusty little 12,000 cubic feet balloon, the same programme, and the same success. This is slightly varied, however, at Kobe, Japan, where his impatient craft fairly breaks away with him, and, soaring high, flies overhead of a man-of-war, and plumps into the water a mile out at sea. But "Smartly" was the word. The ship's crew was beat to quarters, and within one minute a boat was to the rescue. An ascent at Cairo, where he made a parachute descent in sight of the Pyramids and landed in the desert, completed this oriental tour, and home duties necessitated his return to England. Among exploits far too many to enumerate may be mentioned four several occasions when Mr. Percival Spencer has crossed the English Channel.

It fell to the lot of the second son, Arthur, to carry fame into fresh fields. In the year 1897 he visited Australia, taking with him two balloons, one of these being a noble craft of 80,000 cubic feet, considerably larger than any balloon used in England, and the singular fate of this aerial monster is deserving of mention.

Its trial trip in the new country was arranged to take place on Boxing Day in the Melbourne Exhibition ground, and for the lengthy and critical work of inflation the able assistance of British bluejackets was secured. To all appearance, the main difficulties to be provided against were likely to arise simply from a somewhat inadequate supply of gas, and on this account filling commenced as early as 10 a.m. on the morning of the day previous to the exhibition, and was continued till 6 o'clock in the afternoon, by which time the balloon, being about half full, was staved down with sandbags through the night till 4 o'clock the next morning, when the inflation was again proceeded with without hindrance and apparently under favourable conditions. The morning was beautifully fine, warm, brilliant, and still, and so remained until half-past six, when, with startling rapidity, there blew up a sudden squall known in the country as a "Hot Buster," and in two or three minutes' space a terrific wind storm was sweeping the ground. A dozen men, aiding a dead weight of 220 sandbags, endeavoured to control the plunging balloon, but wholly without avail. Men and bags together were lifted clean up in the air on the windward side, and the silk envelope, not yet completely filled, at once escaped from the net and, flying upwards to a height estimated at 10,000 feet, came to earth again ninety miles away in a score of fragments. Nothing daunted, however, Mr. Spencer at once endeavoured to retrieve his fortunes, and started straightway for the gold-mining districts of Ballarat and Bendigo with a hot-air balloon, with which he successfully gave a series of popular exhibitions of parachute descents. Few aeronauts are more consistently reliable than Mr. Arthur Spencer. A few summers ago in this country he was suddenly called upon to give proof of his prowess and presence of mind in a very remarkable manner. It was at an engagement at Reading, where he had been conducting captive ascents throughout the afternoon, and was requested to conclude the evening with a "right away," in which two passengers had agreed to accompany him. The balloon had been hauled down for the last time, when, by some mistake, the engine used for the purpose proceeded to work its pump without previously disconnecting the hauling gear. The consequence of this was that the cable instantly snapped, and in a moment the large balloon, devoid of ballast, grapnel, or other appliances, and with neck still tied, was free, and started skyward.

The inevitable result of this accident must have been that the balloon in a few seconds would rise to a height where the expansion of the imprisoned gas would burst and destroy it. Mr. Spencer, however, was standing near, and, grasping the situation in a moment, caught at the car as it swung upwards, and, getting hold, succeeded in drawing himself up and so climbing into the ring. Quickly as this was done, the balloon was already distended to the point of bursting, and only the promptest release of gas averted catastrophe.

Mr. Stanley Spencer made himself early known to the world by a series of parachute descents, performed from the roof of Olympia. It was a bold and sensational exhibition, and on the expiration of his engagement the young athlete, profiting by home training, felt fully qualified to attempt any aerial feat connected with the profession of an aeronaut. And at this juncture an eminent American cyclist, visiting the father's factory, suggested to Stanley a business tour in South America.

As an extra attraction it was proposed that a young lady parachutist should be one of the company; so, after a few satisfactory trial exhibitions in England, the party made their way to Rio, Brazil. Here an ascent was arranged, and by the day and hour appointed the balloon was successfully inflated with hydrogen, an enormous concourse collected, and the lady performer already seated in the sling. Then a strange mischance happened. By some means, never satisfactorily explained, the young woman, at the moment of release, slipped from her seat, and the balloon, escaping into the air, turned over and fell among the people, who vindictively destroyed it. Then the crowd grew ungovernable, and threatened the lives of the aeronauts, who eventually were, with difficulty, rescued by the soldiery.

This was a bad start; but with a spare balloon a fresh attempt at an ascent was arranged, though, from another cause, with no better success. This time a furious storm arose, before the inflation was completed, and the balloon, carrying away, was torn to ribbons. Yet a third time, with a hot air balloon now, a performance was advertised and successfully carried out; but, immediately after, Mr. Spencer's American friend succumbed to yellow fever, and the young man, being thrown on his own resources, had to fight his own way until his fortunes had been sufficiently restored to return to England.

A few months later he set sail for Canada, where for several months he had a most profitable career, on one occasion only meeting with some difficulty. He was giving an exhibition on Prince Edward's Island, not far from the sea, but on a day so calm that he did not hesitate to ascend. On reaching 3,000 feet, however, he was suddenly caught by a strong land breeze, which, ere he could reach the water, had carried him a mile out to sea, and here he was only rescued after a long interval, during which he had become much exhausted in his attempts to save his parachute from sinking.

Early in 1892 our traveller visited South Africa with a hot air balloon, and, fortune continuing to favour him, he subsequently returned to Canada, and proceeded thence to the United States and Cuba. It was at Havannah that popular enthusiasm in his favour ran so high that he was presented with a medal by the townsfolk. It was from here also that, a little while after, tidings of his own death reached him, together with most gratifying obituary notices. It would seem that, after his departure, an adventurer, attempting to personate him, met with his death.

In November, 1897, he followed his elder brother's footsteps to the East, and exhibited in Calcutta, Singapore, Canton, and also Hong-Kong, where, for the first and only time in his experience, he met with serious accident. He was about to ascend for the ordinary parachute performance with a hot air balloon, which was being held down by about thirty men, one among them being a Chinaman possessed of much excitability and very long finger nails. By means of these latter the man contrived to gouge a considerable hole in the fabric of the balloon. Mr. Spencer, to avoid a disappointment, risked an ascent, and it was not till the balloon had reached 600 feet that the rent developed into a long slit, and so brought about a sudden fall to earth. Alighting on the side of a mountain, Mr. Spencer lay helpless with a broken leg till the arrival of some British bluejackets, who conveyed him to the nearest surgeon, when, after due attention, he was sent home. Other remarkable exploits, which Mr. Stanley Spencer shared with Dr. Berson and with the writer and his daughter, will be recorded later.

CHAPTER XXIII. NEW DEPARTURES IN AERO-STATION.

After Mr. Coxwell's experiments at Aldershot in 1862 the military balloon, as far as England was concerned, remained in abeyance for nine long years, when the Government appointed a Commission to enquire into its utility, and to conduct further experiments. The members of this committee were Colonel Noble, R.E., Sir F. Abel, Captain Lee, R.E., assisted by Captain Elsdale, R.E., and Captain (now Colonel) Templer. Yet another nine years, however, elapsed before much more was heard of this modernised military engine.

But about the beginning of the eighties the Government had become fully alive to the importance of the subject, and Royal Engineers at Woolwich grew busy with balloon manufacture and experiment. Soon "the sky around London became speckled with balloons." The method of making so-called pure hydrogen by passing steam over red-hot iron was fully tested, and for a time gained favour. The apparatus, weighing some three tons, was calculated to be not beyond the carrying powers of three service waggons, while it was capable of generating enough gas to inflate two balloons in twenty-four hours, a single inflation holding good, under favourable circumstances, for a long period. At the Brighton Volunteer Review of 1880, Captain Templer, with nine men, conducted the operations of a captive reconnoitring balloon. This was inflated at the Lewes gas works, and then towed two and a half miles across a river, a railway, and a line of telegraph wires, after which it was let up to a height of 1,500 feet, whence, it was stated, that so good a view was obtained that "every man was clearly seen." Be it remembered, however, that the country was not the South African veldt, and every man was in the striking English uniform of that date.

Just at this juncture came the Egyptian War, and it will be recalled that in the beginning of that war balloons were conspicuous by their absence. The difficulties of reconnaissance were keenly felt and commented on, and among other statements we find the following in the war intelligence of the Times:—

"As the want of a balloon equipment has been mentioned in letters from Egypt, it may be stated that all the War Department balloons remain in store at the Royal Dockyard at Woolwich, but have been recently examined and found perfectly serviceable." An assertion had been made to the effect that the nature of the sand in Egypt would impede the transport of the heavy material necessary for inflation. At last, however, the order came for the despatch of the balloon equipment to the front, and though this arrived long after Tel-el-Kebir, yet it is recorded that the first ascent in real active service in the British Army took place on the 25th of March, 1885, at Suakin, and balloons becoming regarded as an all-important part of the equipment of war, they were sent out in the Bechuanaland Expedition under Sir Charles Warren, the supply of gas being shipped to Cape Town in cylinders.

It was at this period that, according to Mr. Coxwell, Lord Wolseley made ascents at home in a war balloon to form his own personal opinion of their capabilities, and, expressing this opinion to one of his staff, said that had he been able to employ balloons in the earlier stages of the Soudan campaign the affair would not have lasted as many months as it did years. This statement, however, should be read in conjunction with another of the same officer in the "Soldier's Pocket Book," that "in a windy country balloons are useless." In the Boer War the usefulness of the balloon was frequently tested, more particularly during the siege of Ladysmith, when it was deemed of great value in directing the fire of the British artillery, and again in Buller's advance, where the balloon is credited with having located a "death-trap" of the enemy at Spion Kop. Other all-important service was rendered at Magersfontein. The Service balloon principally used was made of goldbeaters' skin, containing about 10,000 cubic feet of hydrogen, which had been produced by the action of sulphuric acid on zinc, and compressed in steel cylinders. A special gas factory was, for the purpose of the campaign, established at Cape Town.

It is here that reference must be made to some of the special work undertaken by Mr. Eric S. Bruce, which dealt with the management of captive balloons under different conditions, and with a system of signalling thus rendered feasible. Mr. Bruce, who, since Major Baden-Powell's retirement from the office, has devoted his best energies as secretary to the advancement of the British Aeronautical Society, was the inventor of the system of electric balloon signalling which he supplied to the British Government, as well as to the Belgian and Italian Governments. This system requires but a very small balloon, made of three or four thicknesses of goldbeaters' skin, measuring from 7 to 10 feet in diameter, and needing only two or three gas cylinders for inflation. Within the balloon, which is sufficiently translucent, are placed several incandescent lamps in metallic circuit, with a source of electricity on the ground.

This source of electricity may consist of batteries of moderate size or a portable hand dynamo. In the circuit is placed an apparatus for making and breaking contact rapidly, and by varying the duration of the flashes in the balloon telegraphic messages may be easily transmitted. To overcome the difficulty of unsteadiness, under circumstances of rough weather, in the captive balloon which carried the glow lamps, Mr. Bruce experimented with guy ropes, and gave a most successful exhibition of their efficiency before military experts at Stamford Bridge grounds, though a stiff wind was blowing at the time.

It must be perfectly obvious, however, that a captive balloon in a wind is greatly at a disadvantage, and to counteract this, attempts have been made in the direction of a combination between the balloon and a kite. This endeavour has been attended with some measure of success in the German army. Mr. Douglas Archibald, in England, was one of the first to advocate the kite balloon. In 1888 he called attention to the unsatisfactory behaviour of captive balloons in variable winds, dropping with every gust and rising again with a lull. In proof he described an expedient of Major Templer's, where an attempt was being made to operate a photographic camera hoisted by two tandem kites. "The balloon," he writes, "went up majestically, and all seemed very satisfactory until a mile of cable had been run out, and the winder locked." It was then that troubles began which threatened the wreckage of the apparatus, and Mr. Archibald, in consequence, strongly recommended a kite balloon at that time. Twelve years later the same able experimentalist, impressed with the splendid work done by kites alone for meteorological purposes at least, allowed that he was quite content to "let the kite balloon go by."

But the German school of aeronauts were doing bigger things than making trials with kite balloons. The German Society for the Promotion of Aerial Navigation, assisted by the Army Balloon Corps, were busy in 1888, when a series of important ascents were commenced. Under the direction of Dr. Assmann, the energetic president of the aeronautical society above named, captive ascents were arranged in connection with free ascents for meteorological purposes, and it was thus practicable to make simultaneous observations at different levels. These experiments, which were largely taken up on the Continent, led to others of yet higher importance, in which the unmanned balloon took a part. But the Continental annals of this date contain one unhappy record of another nature, the recounting of which will, at least, break the monotony attending mere experimental details.

In October, 1893, Captain Charbonnet, an enthusiastic French aeronaut, resolved on spending his honeymoon, with the full consent of his bride, in a prolonged balloon excursion. The start was to be made from Turin, and, the direction of travel lying across the Alps, it was the hope of the voyagers eventually to reach French territory. The ascent was made in perfect safety, as was also the first descent, at the little village of Piobesi, ten miles away. Here a halt was made for the night, and the next morning, when a fresh start was determined on, two young Italians, Signori Botto and Durando, were taken on board as assistants, for the exploit began to assume an appearance of some gravity, and this the more so when storm clouds began brewing. At an altitude of 10,000 feet cross-currents were encountered, and the course becoming obscured the captain descended to near the earth, where he discovered himself to be in dangerous proximity to gaunt mountain peaks. On observing this, he promptly cast out sand so liberally that the balloon rose to a height approaching 20,000 feet, when a rapid descent presently began, and refused to be checked, even with the expenditure of all available ballast.

All the while the earth remained obscured, but, anticipating a fall among the mountains, Captain Charbonnet bade his companions lie down in the car while he endeavoured to catch sight of some landmark; but, quite suddenly, the balloon struck some mountain slope with such force as to throw the captain back into the car with a heavy blow over the eye; then, bounding across a gulley, it struck again and yet again, falling and rebounding between rocky walls, till it settled on a steep and snowy ridge. Darkness was now closing in, and the party, without food or proper shelter, had to pass the night as best they might on the bare spot where they fell, hoping for encouragement with the return of day. But dawn showed them to be on a dangerous peak, 10,000 feet high, whence they must descend by their own unassisted efforts. After a little clambering the captain, who was in a very exhausted state, fell through a hidden crevasse, fracturing his skull sixty feet below. The remaining three struggled on throughout the day, and had to pass a second night on the mountain, this time without covering. On the third day they met with a shepherd, who conducted them with difficulty to the little village of Balme.

This story, by virtue of its romance, finds a place in these pages; but, save for its tragic ending, it hardly stands alone. Ballooning enterprise and adventure were growing every year more and more common on the Continent. In Scandinavia we find the names of Andree, Fraenkal, and Strindberg; in Denmark that of Captain Rambusch. Berlin and Paris had virtually become the chief centres of the development of ballooning as a science. In the former city a chief among aeronauts had arisen in Dr. A. Berson, who, in December, 1894, not only reached 30,000 feet, ascending alone, but at that height sustained himself sufficiently, by inhaling oxygen, to take systematic observations throughout the entire voyage of five hours. The year before, in company with Lieutenant Gross, he barely escaped with his life, owing to tangled ropes getting foul of the valve. Toulet and those who accompanied him lost their lives near Brussels. Later Wolfert and his engineer were killed near Berlin, while Johannsen and Loyal fell into the Sound. Thus ever fresh and more extended enterprise was embarked upon with good fortune and ill. In fact, it had become evident to all that the Continent afforded facilities for the advancement of aerial exploration which could be met with in no other parts of the world, America only excepted. And it was at this period that the expedient of the ballon sonde, or unmanned balloon, was happily thought of. One of these balloons, the "Cirrus," among several trials, rose to a height, self-registered, of 61,000 feet, while a possible greater height has been accorded to it. On one occasion, ascending from Berlin, it fell in Western Russia, on another in Bosnia. Then, in 1896, at the Meteorological Conference at Paris, with Mascart as President, Gustave Hermite, with characteristic ardour, introduced a scheme of national ascents with balloons manned and unmanned, and this scheme was soon put in effect under a commission of famous names — Andree, Assmann, Berson, Besancon, Cailletet, Erk, de Fonvielle, Hergesell, Hermite, Jaubert, Pomotzew (of St. Petersburg), and Rotch (of Boston, Mass.).

In November, 1896, five manned balloons and three unmanned ascended simultaneously from France, Germany, and Russia. The next year saw, with the enterprise of these nations, the co-operation of Austria and Belgium. Messrs. Hermite and Besancon, both French aeronauts, were the first to make practical trial of the method of sounding the upper air by unmanned balloons, and, as a preliminary attempt, dismissed from Paris a number of small balloons, a large proportion of which were recovered, having returned to earth after less than 100 miles' flight. Larger paper balloons were now constructed, capable of carrying simple self-recording instruments, also postcards, which became detached at regular intervals by the burning away of slow match, and thus indicated the path of the balloon. The next attempt was more ambitious, made with a gold-beaters' skin balloon containing 4,000 cubic feet of gas, and carrying automatic instruments of precision. This balloon fell in the Department of the Yonne, and was returned to Paris with the instruments, which remained uninjured, and which indicated that an altitude of 49,000 feet had been reached, and a minimum temperature of -60 degrees encountered. Yet larger balloons of the same nature were then experimented with in Germany, as well as France.

A lack of public support has crippled the attempts of experimentalists in this country, but abroad this method of aerial exploration continues to gain favour.

Distinct from, and supplementing, the records obtained by free balloons, manned or unmanned, are those to be gathered from an aerostat moored to earth. It is here that the captive balloon has done good service to meteorology, as we have shown, but still more so has the high-flying kite. It must long have been recognised that instruments placed on or near the ground are insufficient for meteorological purposes, and, as far back as 1749, we find Dr. Wilson, of Glasgow, employing kites to determine the upper currents, and to carry thermometers into higher strata of the air. Franklin's kite and its application is matter of history. Many since that period made experiments more or less in earnest to obtain atmospheric observations by means of kites, but probably the first in England, at least to obtain satisfactory results, was Mr. Douglas Archibald, who, during the eighties, was successful in obtaining valuable wind measurements, as also other results, including aerial photographs, at varying altitudes up to 1,000 or 1,200 feet. From that period the records of serious and systematic kite flying must be sought in America. Mr. W. A. Eddy was one of the pioneers, and a very serviceable tailless kite, in which the cross-bar is bowed away from the wind, is his invention, and has been much in use. Mr. Eddy established his kite at Blue Hill—the now famous kite observatory—and succeeded in lifting self-recording meteorological instruments to considerable heights. The superiority of readings thus obtained is obvious from the fact that fresh air-streams are constantly playing on the instruments.

A year or two later a totally dissimilar kite was introduced by Mr. Lawrence Hargrave, of Sydney, Australia. This invention, which has proved of the greatest utility and efficiency, would, from its appearance, upset all conventional ideas of what a kite should be, resembling in its simplest form a mere box, minus the back and front. Nevertheless, these kites, in their present form, have carried instruments to heights of upwards of two miles, the restraining line being fine steel piano wire.

But another and most efficient kite, admirably adapted for many most important purposes, is that invented by Major Baden-Powell. The main objects originally aimed at in the construction of this kite related to military operations, such as signalling, photography, and the raising of a man to an elevation for observational purposes. In the opinion of the inventor, who is a practiced aeronaut, a wind of over thirty miles an hour renders a captive balloon useless, while a kite under such conditions should be capable of taking its place in the field. Describing his early experiments, Major, then Captain, Baden-Powell, stated that in 1894, after a number of failures, he succeeded with a hexagonal structure of cambric, stretched on a bamboo framework 36 feet high, in lifting a man—not far, but far enough to prove that his theories were right. Later on, substituting a number of small kites for one big one, he was, on several occasions, raised to a height of 100 feet, and had sent up sand bags, weighing 9 stone, to 300 feet, at which height they remained suspended nearly a whole day.

This form of kite, which has been further developed, has been used in the South African campaign in connection with wireless telegraphy for the taking of photographs at great heights, notably at Modder River, and for other purposes.

It has been claimed that the first well-authenticated occasion of a man being raised by a kite was when at Pirbright Camp a Baden-Powell kite, 30 feet high, flown by two lines, from which a basket was suspended, took a man up to a height of 10 feet. It is only fair, however, to state that it is related that more than fifty years ago a lady was lifted some hundred feet by a great kite constructed by one George Pocock, whose machine was designed for an observatory in war, and also for drawing carriages along highways.

CHAPTER XXIV. ANDREE AND HIS VOYAGES

Among many suggestions, alike important and original, due to Major Baden-Powell, and coming within the field of aeronautics, is one having reference to the use of balloons for geographical research generally and more particularly for the exploration of Egypt, which, in his opinion, is a country possessing many most desirable qualifications on the score of prevailing winds, of suitable base, and of ground adapted for such steering as may be effected with a trail rope. At the Bristol meeting of the British Association the Major thus propounded his method: "I should suggest several balloons, one of about 60,000 cubic feet, and, say, six smaller ones of about 7,000 cubic feet; then, if one gets torn or damaged, the others might remain intact. After a time, when gas is lost, one of the smaller ones could be emptied into the others, and the exhausted envelope discharged as ballast; the smaller balloons would be easier to transport by porters than one big one, and they could be more easily secured on the earth during contrary winds. Over the main balloon a light awning might be rigged to neutralise, as far as possible, the changes of temperature. A lightning conductor to the top of the balloon might be desirable. A large sail would be arranged, and a bifurcated guide rope attached to the end of a horizontal pole would form an efficient means of steering. The car would be boat-shaped and waterproof, so that it could be used for a return journey down a river. Water tanks would be fitted."

The reasonableness of such a scheme is beyond question, even without the working calculations with which it is accompanied; but, ere these words were spoken, one of the most daring explorers that the world has known had begun to put in practice a yet bolder and rasher scheme of his own. The idea of reaching the North Pole by means of balloons appears to have been entertained many years ago. In a curious work, published in Paris in 1863 by Delaville Dedreux, there is a suggestion for reaching the North Pole by an aerostat which should be launched from the nearest accessible point, the calculation being that the distance from such a starting place to the Pole and back again would be only some 1,200 miles, which could be covered in two days, supposing only that there could be found a moderate and favourable wind in each direction. Mr. C. G. Spencer also wrote on the subject, and subsequently Commander Cheyne proposed a method of reaching the Pole by means of triple balloons.

A similar scheme was advocated in yet more serious earnest by M. Hermite in the early eighties.

Some ten years later than this M. S. A. Andree, having obtained sufficient assistance, took up the idea with the determined intention of pushing it to a practical issue. He had already won his spurs as an aeronaut, as may be briefly told. In October, 1893, when making an ascent for scientific purposes, his balloon got carried out over the Baltic. It may have been the strength of the wind that had taken him by surprise; but, there being now no remedy, it was clearly the speed and persistence of the wind that alone could save him. If a chance vessel could not, or would not, "stand by," he must make the coast of Finland or fall in the sea, and several times the fall in the sea seemed imminent as his balloon commenced dropping. This threatened danger induced him to cast away his anchor, after which the verge of the Finland shore was nearly reached, when a change of wind began to carry him along the rocky coast, just as night was setting in.

Recognising his extreme danger, Andree stood on the edge of the car, with a bag of ballast ready for emergencies. He actually passed over an island, on which was a building with a light; but failed to effect a landing, and so fell in the sea on the farther side; but, the balloon presently righting itself, Andree, now greatly exhausted, made his last effort, and as he rose over the next cliff jumped for his life. It was past 7 p.m. when he found himself once again on firm ground, but with a sprained leg and with no one within call. Seeking what shelter he could, he lived out the long night, and, being now scarce able to stand, took off his clothes and waved them for a signal. This signal was not seen, yet shortly a boat put off from an island — the same that he had passed the evening before — and rowed towards him. The boatman overnight had seen a strange sail sweeping over land and sea, and he had come in quest of it, bringing timely succour to the castaway.

Briefly stated, Andree's grand scheme was to convey a suitable balloon, with means for inflating it, as also all necessary equipment, as far towards the Pole as a ship could proceed, and thence, waiting for a favourable wind, to sail by sky until the region of the Pole should be crossed, and some inhabited country reached beyond. The balloon was to be kept near the earth, and steered, as far as this might be practicable, by means of a trail rope. The balloon, which had a capacity of nearly 162,000 cubic feet, was made in Paris, and was provided with a rudder sail and an arrangement whereby the hang of the trail rope could be readily shifted to different positions on the ring. Further, to obviate unnecessary diffusion and loss of gas at the mouth, the balloon was fitted with a lower valve, which would only open at a moderate pressure, namely, that of four inches of water.

All preparations were completed by the summer of 1896, and on June 7th the party embarked at Gothenburg with all necessaries on board, arriving at Spitzbergen on June 21st. Andree, who was to be accompanied on his aerial voyage by two companions, M. Nils Strindberg and Dr. Ekholm, spent some time in selecting a spot that would seem suitable for their momentous start, and this was finally found on Dane's Island, where their cargo was accordingly landed.

The first operation was the erection of a wooden shed, the materials for which they had brought with them, as a protection from the wind. It was a work which entailed some loss of time, after which the gas apparatus had to be got into order, so that, in spite of all efforts, it was the 27th of July before the balloon was inflated and in readiness.

A member of an advance party of an eclipse expedition arriving in Spitzbergen at this period, and paying a visit to Andree for the purpose of taking him letters, wrote:—"We watched him deal out the letters to his men. They are all volunteers and include seven sea captains, a lawyer, and other people some forty in all. Andree chaffed each man to whom he gave a letter, and all were as merry as crickets over the business.... We spent our time in watching preparations. The vaseline (for soaking the guide ropes) caught fire to-day, but, luckily no rope was in the pot."

But the wind as yet was contrary, and day after day passed without any shift to a favourable quarter, until the captain of the ship which had conveyed them was compelled to bring matters to an issue by saying that they must return home without delay if he was to avoid getting frozen in for the winter. The balloon had now remained inflated for twenty-one days, and Dr. Ekholm, calculating that the leakage of gas amounted to nearly 1 per cent. per day, became distrustful of the capability of such a vessel to cope with such a voyage as had been aimed at. The party had now no choice but to return home with their balloon, leaving, however, the shed and gas-generating apparatus for another occasion.

This occasion came the following summer, when the dauntless explorers returned to their task, leaving Gothenburg on May 28th, 1897, in a vessel lent by the King of Sweden, and reaching Dane's Island on the 30th of the same month. Dr.

Ekholm had retired from the enterprise, but in his place were two volunteers, Messrs. Frankel and Svedenborg, the latter as "odd man," to fill the place of any of the other three who might be prevented from making the final venture.

It was found that the shed had suffered during the winter, and some time was spent in making the repairs and needful preparation, so that the month of June was half over before all was in readiness for the inflation. This operation was then accomplished in four days, and by midnight of June 22nd the balloon was at her moorings, full and in readiness; but, as in the previous year, the wind was contrary, and remained so for nearly three weeks. This, of course, was a less serious matter, inasmuch as the voyagers were a month earlier with their preparation, but so long a delay must needs have told prejudicially against the buoyancy of the balloon, and Andree is hardly to be blamed for having, in the end, committed himself to a wind that was not wholly favourable.

The wind, if entirely from the right direction, should have been due south, but on July 11th it had veered to a direction somewhat west of south, and Andree, tolerating no further delay, seized this as his best opportunity, and with a wind "whistling through the woodwork of the shed and flapping the canvas," accompanied by

Frankel and Svedenborg, started on his ill-fated voyage.

A telegram which Andree wrote for the Press at that epoch ran thus: — "At this moment, 2.30 p.m., we are ready to start. We shall probably be driven in a north-north-easterly direction."

On July 22nd a carrier pigeon was recovered by the fishing boat Alken between North Cape, Spitzbergen, and Seven Islands, bearing a message, "July 13th, 12.30 p.m., 82 degrees 2 minutes north lat., 15 degrees 5 minutes east long. Good journey eastward. All goes well on board. Andree."

Not till August 31st was there picked up in the Arctic zone a buoy, which is preserved in the Museum of Stockholm. It bears the message, "Buoy No. 4. First to be thrown out. 11th July, 10 p.m., Greenwich mean time. All well up till now. We are pursuing our course at an altitude of about 250 metres Direction at first northerly 10 degrees east; later; northerly 45 degrees east. Four carrier pigeons were despatched at 5.40 p.m. They flew westwards. We are now above the ice, which is very cut up in all directions. Weather splendid. In excellent spirits. — Andree, Svedenborg, Frankel. (Postscript later on.) Above the clouds, 7.45, Greenwich mean time."

According to Reuter, the Anthropological and Geological Society at Stockholm received the following telegram from a ship owner at Mandal:—"Captain Hueland, of the steamship Vaagen who arrived there on Monday morning, reports that when off Kola Fjord, Iceland, in 65 degrees 34 minutes north lat., 21 degrees 28 minutes west long., on May 14th he found a drifting buoy, marked 'No. 7.' Inside the buoy was a capsule marked 'Andree's Polar Expedition,' containing a slip of paper, on which was given the following: 'Drifting Buoy No. 7. This buoy was thrown out from Andree's balloon on July 11th 1897, 10.55 p.m., Greenwich mean time, 82 degrees north lat., 25 degrees east lon. We are at an altitude of 600 metres. All well.—Andree, Svedenborg, Frankel.'"

Commenting on the first message, Mr. Percival Spencer says:—"I cannot place reliance upon the accuracy of either the date or else the lat. and long. given, as I am confident that the balloon would have travelled a greater distance in two days." It should be noted that Dane's Island lies in 79 degrees 30 minutes north lat. and 10 degrees 10 minutes east long.

Mr. Spencer's opinion, carefully considered and expressed eighteen months afterwards, will be read with real interest:—

"The distance from Dane's Island to the Pole is about 750 miles, and to Alaska on the other side about 1,500 miles. The course of the balloon, however, was not direct to the Pole, but towards Franz Josef Land (about 600 miles) and to the Siberian coast (another 800 miles). Judging from the description of the wind at the start, and comparing it with my own ballooning experience, I estimate its speed as 40 miles per hour, and it will, therefore, be evident that a distance of 2,000 miles would be covered in 50 hours, that is two days and two hours after the start. I regard all theories as to the balloon being capable of remaining in the air for a month as illusory. No free balloon has ever remained aloft for more than 36 hours, but with the favourable conditions at the northern regions (where the sun does not set and where the temperature remains equable) a balloon might remain in the air for double the length of time which I consider ample for the purpose of Polar exploration."

A record of the direction of the wind was made after Andree's departure, and proved that there was a fluctuation in direction from S.W. to N.W., indicating that the voyagers may have been borne across towards Siberia. This, however, can be but surmise. All aeronauts of experience know that it is an exceedingly difficult manoeuvre to keep a trail rope dragging on the ground if it is desirable to prevent contact with the earth on the one hand, or on the other to avoid loss of gas. A slight increase of temperature or drying off of condensed moisture may—indeed, is sure to after a while—lift the rope off the ground, in which case the balloon, rising into upper levels, may be borne away on currents which may be of almost any direction, and of which the observer below may know nothing. As to the actual divergence from the wind's direction which a trail rope and side sail might be hoped to effect, it may be confidently stated that, notwithstanding some wonderful accounts that have gone abroad, it must not be relied on as commonly amounting to much more than one or, at the most, two points.

Although it is to be feared that trustworthy information as to the ultimate destination of Andree's balloon may never be gained, yet we may safely state that his ever famous, though regrettable, voyage was the longest in duration ever attained. At the end of 48 hours his vessel would seem to have been still well up and going strong. The only other previous voyage that had in duration of travel approached this record was that made by M. Mallet, in 1892, and maintained for 36 hours. Next we may mention that of M. Herve, in 1886, occupying 24 1/2 hours, which feat, however, was almost equal led by the great Leipzig balloon in 1897, which, with eight people in the car, remained up for 24 1/4 hours, and did not touch earth till 1,032 miles had been traversed.

The fabric of Andree's balloon may not be considered to have been the best for such an exceptional purpose. Dismissing considerations of cost, goldbeaters' skin would doubtless have been more suitable. The military balloons at Aldershot are made of this, and one such balloon has been known to remain inflated for three months with very little loss. It is conceivable, therefore, that the chances of the voyagers, whose ultimate safety depended so largely upon the staying power of their aerial vessel, might have been considerably increased.

One other expedient, wholly impracticable, but often seriously discussed, may be briefly referred to, namely, the idea of taking up apparatus for pumping gas into metal receivers as the voyage proceeds, in order to raise or lower a balloon, and in this way to prolong its life. Mr. Wenham has investigated the point with his usual painstaking care, and reduced its absurdity to a simple calculation, which should serve to banish for good such a mere extravagant theory.

Suppose, he says, the gas were compressed to one-twentieth part of its bulk, which would mean a pressure within its receiver of 300 lbs. per square inch, and that each receiver had a capacity of 1 cubic foot, while for safety sake it was made of steel plates one-twentieth of an inch thick, then each receiver would weigh 10 lbs., and to liberate 1,000 feet clearly a weight of 500 lbs. would have to be taken up. Now, when it is considered that 1,000 cubic feet of hydrogen will only lift 72 lbs., the scheme begins to look hope less enough. But when the question of the pumping apparatus, to be worked by hand, is contemplated the difficulties introduced become yet more insuperable. The only feasible suggestion with respect the use of compressed gas is that of taking on board charged cylinders under high pressure, which, after being discharged to supply the leakage of the balloon could, in an uninhabited country, be cast out as ballast last. It will need no pointing out, however, that such an idea would be practically as futile as another which has gravely been recommended, namely, that of heating the gas of the balloon by a Davy lamp, so as to increase its buoyancy at will. Major Baden-Powell has aptly described this as resembling "an attempt to warm a large hall with a small spirit lamp."

In any future attempt to reach the Pole by balloon it is not unreasonable to suppose that wireless telegraphy will be put in practice to maintain communication with the base. The writer's personal experience of the possibilities afforded by this mode of communication, yet in its infancy, will be given.

CHAPTER XXV. THE MODERN AIRSHIP – IN SEARCH OF THE LEONIDS.

In the autumn of 1898 the aeronautical world was interested to hear that a young Brazilian, M. Santos Dumont, had completed a somewhat novel dirigible balloon, cylindrical in shape, with conical ends, 83 feet long by 12 feet in diameter, holding 6,500 cubic feet of gas, and having a small compensating balloon of 880 cubic feet capacity. For a net was substituted a simple contrivance, consisting of two side pockets, running the length of the balloon, and containing battens of wood, to which were affixed the suspension cords, bands being also sewn over the upper part of the balloon connecting the two pockets. The most important novelty, however, was the introduction of a small petroleum motor similar to those used for motor tricycles.

The inventor ascended in this balloon, inflated with pure hydrogen, from the Jardin d'Acclimatation, Paris, and circled several times round the large captive balloon in the Gardens, after which, moving towards the Bois de Boulogne, he made several sweeps of 100 yards radius. Then the pump of the compensator caused the engine to stop, and the machine, partially collapsing, fell to the ground. Santos Dumont was somewhat shaken, but announced his intention of making other trials. In this bold and successful attempt there was clear indication of a fresh phase in the construction of the airship, consisting in the happy adoption of the modern type of petroleum motor. Two other hying machines were heard of about this date, one by Professor Giampietre, of Pavia, cigar-shaped, driven by screws, and rigged with masts and sails. The other, which had been constructed and tested in strict privacy, was the invention of a French engineer, M. Ader, and was imagined to imitate the essential structure of a bird. Two steam motors of 20-horse power supplied the power. It was started by being run on the ground on small wheels attached to it, and it was claimed that before a breakdown occurred the machine had actually raised itself into the air.

Of Santos Dumont the world was presently to know more, and the same must be said of another inventor, Dr. Barton, of Beckenham, who shortly completed an airship model carrying aeroplanes and operated by clockwork. In an early experiment this model travelled four miles in twenty-three minutes.

But another airship, a true leviathan, had been growing into stately and graceful proportions on the shores of the Bodenzee in Wurtemberg, and was already on the eve of completion. Count

Zeppelin, a lieut.-general in the German Army, who had seen service in the Franco-German War, had for some years devoted his fortune and energy to the practical study of aerial navigation, and had prosecuted experiments on a large scale. Eventually, having formed a company with a large capital, he was enabled to construct an airship which in size has been compared to a British man-of-war. Cigar-shaped, its length was no less than 420 feet, and diameter 40 feet, while its weight amounted to no more than 7,250 lbs. The framework, which for lightness had been made of aluminium, was, with the object of preventing all the gas collecting at one end of its elongated form, subdivided into seventeen compartments, each of these compartments containing a completely fitted gas balloon, made of oiled cotton and marvellously gas tight. A steering apparatus was placed both fore and aft, and at a safe distance below the main structure were fixed, also forward and aft, on aluminium platforms, two Daimler motor engines of 16-horse power, working aluminium propellers of four blades at the rate of 1,000 revolutions a minute. Finally, firmly attached to the inner framework by rods of aluminium, were two cars of the same metal, furnished with buffer springs to break the force of a fall. The trial trip was not made till the summer following—June, 1900—and, in the meanwhile, experiments had gone forward with another mode of flight, terminating, unhappily, in the death of one of the most expert and ingenious of mechanical aeronauts.

Mr. Percy S. Pilcher, now thirty-three years of age, having received his early training in the Navy, retired from the Service to become a civil engineer, and had been for some time a partner in the firm of Wilson and Pilcher. For four or five years he had been experimenting in soaring flight, using a Lilienthal machine, which he improved to suit his own methods. Among these was the device of rising off the ground by being rapidly towed by a line against the wind.

At the end of September he gave an exhibition at Stamford Park before Lord Bray and a select party of friends—this in spite of an unsuitable afternoon of unsteady wind and occasional showers. A long towing line was provided, which, being passed round pulley blocks and dragged by a couple of horses, was capable of being hauled in at high speed. The first trial, though ending in an accident, was eminently satisfactory. The apparatus, running against the wind, had risen some distance, when the line broke, yet the inventor descended slowly and safely with outstretched wings. The next trial also commenced well, with an easy rise to a height of some thirty feet. At that point, however, the tail broke with a snap, and the machine, pitching over, fell a complete wreck. Mr. Pilcher was found insensible, with his thigh broken, and though no other serious injury was apparent, he succumbed two days afterwards without recovering consciousness. It was surmised that shrinkage of the canvas of the tail, through getting wet, had strained and broken its bamboo stretcher.

This autumn died Gaston Tissandier, at the age of fifty-six; and in the month of December, at a ripe old age, while still in full possession of intellectual vigour, Mr. Coxwell somewhat suddenly passed away. Always keenly interested in the progress of aeronautics; he had but recently, in a letter to the Standard, proposed a well-considered and practical method of employing Montgolfier reconnoitring balloons, portable, readily inflated, and especially suited to the war in South Africa. Perhaps the last letters of a private nature penned by Mr. Coxwell were to the writer and his daughter, full of friendly and valuable suggestion, and more particularly commenting on a recent scientific aerial voyage, which proved to be not only sensational, but established a record in English ballooning.

The great train of the November meteors, known as the Leonids, which at regular periods of thirty-three years had in the past encountered the earth's atmosphere, was due, and over-due. The cause of this, and of their finally eluding observation, need only be very briefly touched on here. The actual meteoric train is known to travel in an elongated ellipse, the far end of which lies near the confines of the solar system, while at a point near the hither end the earth's orbit runs slantingly athwart it, forming, as it were, a level crossing common to the two orbits, the earth taking some five or six hours in transit. Calculation shows that the meteor train is to be expected at this crossing every thirty-three and a third years, while the train is extended to such an enormous length—taking more than a year to draw clear—that the earth must needs encounter it ere it gets by, possibly even two years running. There could be no absolute certainty about the exact year, nor the exact night when the earth and the meteors would foregather, owing to the uncertain disturbance which the latter must suffer from the pull of the planetary bodies in the long journey out and home again among them. As is now known, this disturbing effect had actually dispersed the train.

The shower, which was well seen in 1866, was pretty confidently expected in 1899, and to guard against the mischance of cloudy weather, it was arranged that the writer should, on behalf of the Times newspaper, make an ascent on the right night to secure observations. Moreover, it was arranged that he should have, as chief assistant, his own daughter, an enthusiastic lady aeronaut, who had also taken part in previous astronomical work.

Unfortunately there were two nights, those of November 14th and 15th, when the expected shower seemed equally probable, and, taking counsel with the best authorities in the astronomical world, it seemed that the only course to avoid disappointment would be to have a balloon filled and moored in readiness for an immediate start, either on the first night or on the second.

This settled the matter from the astronomical side, but there was the aeronautical side also to be considered. A balloon of 56,000 cubic feet capacity was the largest available for the occasion, and a night ascent with three passengers and instruments would need plenty of lifting power to meet chance emergencies. Thus it seemed that a possible delay of forty-eight hours might entail a greater leakage of gas than could be afforded.

The leakage might be expected chiefly to occur at the valve in the head of the balloon, it being extremely difficult to render any form of mechanical valve gas tight, however carefully its joints be stopped with luting. On this account, therefore, it was determined that the balloon should be fitted with what is known as a solid or rending valve, consisting simply of balloon fabric tied hard and fast over the entire upper outlet, after the fashion of a jam pot cover. The outlet itself was a gaping hole of over 2 feet across; but by the time its covering had been carefully varnished over all leakage was sufficiently prevented, the one drawback to this method being the fact that the liberation of gas now admitted of no regulation. Pulling the valve line would simply mean opening the entire wide aperture, which could in no way be closed again.

The management of such a valve consists in allowing the balloon to sink spontaneously earthwards, and when it has settled near the ground, having chosen a desirable landing place, to tear open the so-called valve once and for all.

This expedient, dictated by necessity, seeming sufficient for the purpose at hand, preparations were proceeded with, and, under the management of Mr. Stanley Spencer, who agreed to act as aeronaut, a large balloon, with solid valve, was brought down to Newbury gas works on November 14th, and, being inflated during the afternoon, was full and made snug by sundown. But as the meteor radiant would not be well above the horizon till after midnight, the aeronautical party retired for refreshment, and subsequently for rest, when, as the night wore on, it became evident that, though the sky remained clear, there would be no meteor display that night. The next day was overcast, and by nightfall hopelessly so, the clouds ever thickening, with absence of wind or any indication which might give promise of a change. Thus by midnight it became impossible to tell whether any display were in progress or not. Under these circumstances, it might have been difficult to decide when to make the start with the best show of reason. Clearly too early a start could not subsequently be rectified; the balloon, once off, could not come back again; while, once liberated, it would be highly unwise for it to remain aloft and hidden by clouds for more than some two hours, lest it should be carried out to sea.

Happily the right decision under these circumstances was perfectly clear. Other things being equal, the best time would be about 4 a.m., by which period the moon, then near the full, would be getting low, and the two hours of darkness left would afford the best seeing. Leaving, then, an efficient outlook on the balloon ground, the party enjoyed for some hours the entertainment offered them by the Newbury Guildhall Club, and at 4 a.m. taking their seats in the car, sailed up into the calm chilly air of the November night.

But the chilliness did not last for long. A height of 1,500 feet was read by the Davy lamp, and then we entered fog—warm, wetting fog, through which the balloon would make no progress in spite of a prodigal discharge of sand. The fact was that the balloon, which had become chilled through the night hours, was gathering a great weight of moisture from condensation on its surface, and when, at last, the whole depth of the cloud, 1,500 feet, had been penetrated, the chill of the upper air crippled the balloon and sent her plunging down again into the mist, necessitating yet further expenditure of sand, which by this time had amounted to no less than 3 1/2 cwt. in twenty minutes. And then at last we reached our level, a region on the upper margin of the cloud floor, where evaporation reduced the temperature, that had recently been that of greenhouse warmth, to intense cold.

That evaporation was going on around us on a gigantic scale was made very manifest. The surface of the vast cloud floor below us was in a perfect turmoil, like that of a troubled sea. If the cloud surface could be compared to anything on earth it most resembled sea where waves are running mountains high. At one moment we should be sailing over a trough, wide and deep below us, the next a mighty billow would toss itself aloft and vanish utterly into space. Everywhere wreaths of mist with ragged fringes were withering away into empty air, and, more remarkable yet, was the conflict of wind which sent the cloud wrack flying simply in all directions.

For two hours now there was opportunity for observing at leisure all that could be made of the falling meteors. There were a few, and these, owing to our clear, elevated region, were exceptionally bright. The majority, too, were true Leonids, issuing from the radiant point in the "Sickle," but these were not more numerous than may be counted on that night in any year, and served to emphasise the fact that no real display was in progress. The outlook was maintained, and careful notes made for two hours, at the end of which time the dawn began to break, the stars went in, and we were ready to pack up and come down.

But the point was that we were not coming down. We were at that time, 6 a.m., 4,000 feet high, and it needs no pointing out that at such an altitude it would have been madness to tear open our huge rending valve, thus emptying the balloon of gas. It may also be unnecessary to point out that in an ordinary afternoon ascent such a valve would be perfectly satisfactory, for under these circumstances the sun presently must go down, the air must grow chill, and the balloon must come earthward, allowing of an easy descent until a safe and suitable opportunity for rending the valve occurred; but now we knew that conditions were reversed, and that the sun was just going to rise.

And then it was we realised that we were caught in a trap. From that moment it was painfully evident that we were powerless to act, and were at the mercy of circumstances. By this time the light was strong, and, being well above the tossing billows of mist, we commanded an extended view on every side, which revealed, however, only the upper unbroken surface of the dense cloud canopy that lay over all the British Isles. We could only make a rough guess as to our probable locality. We knew that our course at starting lay towards the west, and if we were maintaining that course a travel of scarcely more than sixty miles would carry us out to the open sea. We had already been aloft for two hours, and as we were at an altitude at which fast upper currents are commonly met with, it was high time that, for safety, we should be coming down; yet it was morally certain that it would be now many hours before our balloon would commence to descend of its own accord by sheer slow leakage of gas, by which time, beyond all reasonable doubt, we must be carried far out over the Atlantic. All we could do was to listen intently for any sounds that might reach us from earth, and assure us that we were still over the land; and for a length of time such sounds were vouchsafed us — the bark of a dog, the lowing of cattle, the ringing trot of a horse on some hard road far down.

And then, as we were expecting, the sun climbed up into an unsullied sky, and, mounting by leaps and bounds, we watched the cloud floor receding beneath us. The effect was extremely beautiful. A description written to the Times the next morning, while the impression was still fresh, and from notes made at this period, ran thus: — "Away to an infinitely distant horizon stretched rolling billows of snowy whiteness, broken up here and there into seeming icefields, with huge fantastic hummocks. Elsewhere domes and spires reared themselves above the general surface, or an isolated Matterhorn towered into space. In some quarters it was impossible to look without the conviction that we actually beheld the outline of lofty cliffs overhanging a none too distant sea." Shortly we began to hear loud reports overhead, resembling small explosions, and we knew what these were—the moist, shrunken netting was giving out under the hot sun and yielding now and again with sudden release to the rapidly expanding gas. It was, therefore, with grave concern, but with no surprise, that when we next turned to the aneroid we found the index pointing to 9,000 feet, and still moving upwards.

Hour after hour passed by, and, sounds having ceased to reach us, it remains uncertain whether or no we were actually carried out to sea and headed back again by contrary currents, an experience with which aeronauts, including the writer, have been familiar; but, at length, there was borne up to us the distant sound of heavy hammers and of frequent trains, from which we gathered that we were probably over Bristol, and it was then that the thought occurred to my daughter that we might possibly communicate with those below with a view to succour. This led to our writing the following message many times over on blank telegraph forms and casting them down:—"Urgent. Large balloon from Newbury travelling overhead above the clouds. Cannot descend. Telegraph to sea coast (coast-guards) to be ready to rescue.—Bacon and Spencer."

While thus occupied we caught the sound of waves, and the shriek of a ship's siren. We were crossing a reach of the Severn, and most of our missives probably fell in the sea. But over the estuary there must have been a cold upper current blowing, which crippled our balloon, for the aneroid presently told of a fall of 2,000 feet. It was now past noon, and to us the turn of the tide was come. Very slowly, and with strange fluctuations, the balloon crept down till it reached and became enveloped in the cloud below, and then the end was near. The actual descent occupied nearly two hours, and affords a curious study in aerostation. The details of the balloon's dying struggles and of our own rough descent, entailing the fracture of my daughter's arm, are told in another volume.{*}

We fell near Neath, Glamorganshire, only one and a half miles short of the sea, completing a voyage which is a record in English ballooning — ten hours from start to finish.

* "By Land and Sky," by the Author.

CHAPTER XXVI. RECENT AERONAUTICAL EVENTS.

The first trial of the Zeppelin air ship was arranged to take place on June 30th, 1900, a day which, from absence of wind, was eminently well suited for the purpose; but the inflation proved too slow a process, and operations were postponed to the morrow. The morrow, however, was somewhat windy, causing delay, and by the time all was in readiness darkness had set in and the start was once more postponed. On the evening of the third day the monster craft was skilfully and successfully manoeuvred, and, rising with a very light wind, got fairly away, carrying Count Zeppelin and four other persons in the two cars. Drifting with the wind, it attained a height of some 800 or 900 feet, at which point the steering apparatus being brought into play it circled round and faced the wind, when it remained stationary. But not for long. Shortly it began to descend and, sinking gradually, gracefully, and in perfect safety, in about nine minutes it reached and rested on the water, when it was towed home.

A little later in the month, July, another trial was made, when a wind was blowing estimated at sixteen miles an hour. As on the previous occasion, the direct influence of the sun was avoided by waiting till evening hours. It ascended at 8 p.m., and the engines getting to work it made a slow progress of about two miles an hour against the wind for about 3 1/2 miles, when one of the rudders gave way, and the machine was obliged to descend.

On the evening of October 24th of the same year, in very calm weather and with better hope, another ascent was made. On this occasion, however, success was frustrated by one of the rear rudders getting foul of the gear, followed by the escape of gas from one of the balloons.

Another and more successful trial took place in the same month, again in calm atmosphere. Inferior gas was employed, and it would appear that the vessel had not sufficient buoyancy. It remained aloft for a period of twenty minutes, during which it proved perfectly manageable, making a graceful journey out and home, and returning close to its point of departure. This magnificent air ship, the result of twenty years of experiment, has since been abandoned and broken up; yet the sacrifice has not been without result. Over and above the stimulus which Count Zeppelin's great endeavour has given to the aeronautical world, two special triumphs are his. He has shown balloonists how to make a perfectly gas-tight material, and has raised powerful petroleum motors in a balloon with safety.

In the early part of 1900 it was announced that a member of the Paris Aero Club, who at the time withheld his name (M. Deutsch) offered a prize of 100,000 francs to the aeronaut who, either in a balloon or flying machine, starting from the grounds of the Aero Club at Longchamps, would make a journey round the Eiffel Tower, returning to the starting place within half an hour. The donor would withdraw his prize if not won within five years, and in the meanwhile would pay 4,000 francs annually towards the encouragement of worthy experimenters.

It was from this time that flying machines in great variety and goodly number began to be heard of, if not actually seen. One of the earliest to be announced in the Press was a machine invented by the Russian, Feedoroff, and the Frenchman, Dupont. Dr. Danilewsky came forward with a flying machine combining balloon and aeroplane, the steering of which would be worked like a velocipede by the feet of the aeronaut.

Mr. P. Y. Alexander, of Bath, who had long been an enthusiastic balloonist, and who had devoted a vast amount of pains, originality, and engineering skill to the pursuit of aeronautics, was at this time giving much attention to the flying machine, and was, indeed, one of the assistants in the first successful launching of the Zeppelin airship. In concert with Mr. W. G. Walker, A.M.I.C.E., Mr. Alexander carried out some valuable and exhaustive experiments on the lifting power of air propellers, 30 feet in diameter, driven by a portable engine. The results, which were of a purely technical nature, have been embodied in a carefully compiled memoir.

An air ship now appeared, invented by M. Rose, consisting of two elongated vessels filled with gas, and carrying the working gear and car between them. The machine was intentionally made heavier than air, and was operated by a petrol motor of 12-horse power.

It was now that announcements began to be made to the effect that, next to the Zeppelin air ship, M. Santos Dumont's balloon was probably attracting most of the attention of experts. The account given of this air vessel by the Daily Express was somewhat startling. The balloon proper was compared to a large torpedo. Three feet beneath this hangs the gasoline motor which is to supply the power. The propeller is 12 feet in diameter, and is revolved so rapidly by the motor that the engine frequently gets red hot. The only accommodation for the traveller is a little bicycle seat, from which the aeronaut will direct his motor and steering gear by means of treadles. Then the inclination or declination of his machine must be noted on the spirit level at his side, and the 200 odd pounds of ballast must be regulated as the course requires.

A more detailed account of this navigable balloon was furnished by a member of the Paris Aero Club. From this authority we learn that the capacity of the balloon was 10,700 cubic feet. It contained an inner balloon and an air fan, the function of which was to maintain the shape of the balloon when meeting the wind, and the whole was operated by a 10-horse power motor capable of working the screw at 100 revolutions per minute.

But before the aerial exploits of Santos Dumont had become famous, balloons had again claimed public attention. On August 1st Captain Spelterini, with two companions, taking a balloon and 180 cylinders of hydrogen to the top of the Rigi and ascending thence, pursued a north-east course, across extensive and beautiful tracts of icefield and mountain fastnesses unvisited by men. The descent, which was difficult and critical, was happily manoeuvred. This took place on the Gnuetseven, a peak over 5,000 feet high, the pla-

teau on which the voyagers landed being described as only 50 yards square, surrounded by pre-
cipices.

On the 10th of September following the writer
was fortunate in carrying out some wireless telegraphy experiments in a balloon, the success of
which is entirely due to the unrivalled skill of Mr.
Nevil Maskelyne, F.R.A.S., and to his clever
adaptation of the special apparatus of his own
invention to the exigencies of a free balloon. The
occasion was the garden party at the Bradford
meeting of the British Association, Admiral Sir
Edmund Fremantle taking part in the voyage,
with Mr. Percival Spencer in charge. The experiment was to include the firing of a mine in the
grounds two minutes after the balloon had left,
and this item was entirely successful. The main
idea was to attempt to establish communication
between a base and a free balloon retreating
through space at a height beyond practicable gun
shot. The wind was fast and squally, and the unavoidable rough jolting which the car received at
the start put the transmitting instrument out of
action. The messages, however, which were sent
from the grounds at Lister Park were received
and watched by the occupants of the car up to a
distance of twenty miles, at which point the voyage terminated.

On September 30th, and also on October 9th, of this year, took place two principal balloon races from Vincennes in connection with the Paris Exposition. In the first race, among those who competed were M. Jacques Faure, the Count de la Vaulx, and M. Jacques Balsan. The Count was the winner, reaching Wocawek, in Russian Poland, a travel of 706 miles, in 21 hours 34 minutes. M. Balsan was second, descending near Dantzig in East Prussia, 757 miles, in 22 hours. M. Jacques Faure reached Mamlitz, in East Prussia, a distance of 753 miles.

In the final race the Count de la Vaulx made a record voyage of 1,193 miles, reaching Korosticheff, in Russia, in 35 hours 45 minutes, attaining a maximum altitude of 18,810 feet. M. J. Balsan reached a greater height, namely, 21,582 feet, travelling to Rodom, in Russia, a distance of 843 miles, in 27 hours 25 minutes.

Some phenomenal altitudes were attained at this time. In September, 1898, Dr. Berson, of Berlin, ascended from the Crystal Palace in a balloon inflated with hydrogen, under the management of Mr. Stanley Spencer, oxygen being an essential part of the equipment. The start was made at 5 p.m., and the balloon at first drifted south-east, out over the mouth of the Thames, until at an altitude of 10,000 feet an upper current changed the course to southwest, the balloon mounting rapidly till 23,000 feet was reached, at which height the coast of France was plainly seen. At 25,000 feet both voyagers were gasping, and compelled to inhale oxygen. At 27,500 feet, only four bags of ballast being left, the descent was commenced, and a safe landing was effected at Romford.

Subsequently Dr. Berson, in company with Dr. Suring, ascending from Berlin, attained an altitude of 34,000 feet. At 30,000 feet the aeronauts were inhaling oxygen, and before reaching their highest point both had for a considerable time remained unconscious.

In 1901 a new aeroplane flying machine began to attract attention, the invention of Herr Kress. A novel feature of the machine was a device to render it of avail for Arctic travel. In shape it might be compared to an iceboat with two keels and a long stem, the keels being adapted to run on ice or snow, while the boat would float on water. Power was to be derived from a petrol motor.

At the same period M. Henry Sutor was busy on Lake Constance with an air ship designed also to float on water. Then Mr. Buchanan followed with a fish-shaped vessel, one of the most important specialities of which consisted in side propellers, the surfaces of which were roughened with minute diagonal grooves to effect a greater grip on the air.

No less original was the air ship, 100 feet long, and carrying 18,000 cubic feet of gas, which Mr. W. Beedle was engaged upon. In this machine, besides the propellers for controlling the horizontal motion, there was one to regulate vertical motion, with a view of obviating expenditure of gas or ballast.

But by this time M. Santos Dumont, pursuing his hobby with unparalleled perseverance, had built in succession no less than six air ships, meeting with no mean success, profiting by every lesson taught by failures, and making light of all accidents, great or small. On July 15th, 1901, he made a famous try for the Deutsch prize in a cigar-shaped balloon, 110 feet long, 19,000 cubic feet capacity, carrying a Daimler oil motor of 15-horse power. The day was not favourable, but, starting from the Parc d'Aerostation, he was abreast of the Eiffel Tower in thirteen minutes, circling round which, and battling against a head wind, he reached the grounds of the Aero Club in 41 minutes from the start, or 11 minutes late by the conditions of the prize. A cylinder had broken down, and the balance of the vessel had become upset.

Within a fortnight—July 29th—in favourable weather, he made another flight, lasting fifteen minutes, at the end of which he had returned to his starting ground. Then on August 8th a more momentous attempt came off. Sailing up with a rapid ascent, and flying with the wind, Santos Dumont covered the distance to the Tower in five minutes only, and gracefully swung round; but, immediately after, the wind played havoc, slowing down the motor, at the same time damaging the balloon, and causing an escape of gas. On this Santos Dumont, ascending higher into the sky, quitted the car, and climbed along the keel to inspect, and, if possible, rectify the motor, but with little success. The balloon was emptying, and the machine pitched badly, till a further rent occurred, when it commenced falling hopelessly and with a speed momentarily increasing.

Slanting over a roof, the balloon caught a chimney and tore asunder; but the wreck, also catching, held fast, while the car hung helplessly down a blank wall. In this perilous predicament great coolness and agility alone averted disaster, till firemen were able to come to the rescue.

The air ship was damaged beyond repair, but by September 6th another was completed, and on trial appeared to work well until, while travelling at speed, it was brought up and badly strained by the trail rope catching in trees.

Early in the next month the young Brazilian was aloft again, with weather conditions entirely in his favour; but again certain minor mishaps prevented his next struggle for the prize, which did not take place till the 19th. On this day a light cross wind was blowing, not sufficient, however, seriously to influence the first stage of the time race, and the outward journey was accomplished with a direct flight in nine minutes. On rounding the tower, however, the wind began to tell prejudicially, and the propeller became deranged. On this, letting his vessel fall off from the wind, Santos Dumont crawled along the framework till he reached the motor, which he succeeded in again setting in working order, though not without a delay of several minutes and some loss of ground. From that point the return journey was accomplished in eight minutes, and the race was, at the time, declared lost by 40 seconds only.

The most important and novel feature in the air ships constructed by Santos Dumont was the internal ballonet, inflated automatically by a ventilator, the expedient being designed to preserve the shape of the main balloon itself while meeting the wind. On the whole, it answered well, and took the place of the heavy wire cage used by Zeppelin.

M. de Fonvielle, commenting on the achievements of Santos Dumont, wrote:—"It does not appear that he has navigated his balloon against more than very light winds, but in his machinery he has shown such attention to detail that it may reasonably be expected that if he continues to increase his motive power he will, ere long, exceed past performances."

Mr. Chanute has a further word to say about the possibility of making balloons navigable. He considers that their size will have to be great to the verge of impracticability and the power of the motor enormous in proportion to its weight. As to flying machines, properly so called, he calculates the best that has been done to be the sustaining of from 27 lbs. to 55 lbs. per horse power by impact upon the air. But Mr. Chanute also argues that the equilibrium is of prime importance, and on this point there could scarcely be a greater authority. No one of living men has given more attention to the problem of "soaring," and it is stated that he has had about a thousand "slides" made by assistants, with different types of machine, and all without the slightest accident.

steadily into the calm sky. The Daily Mail gives the following account of what ensued:—

"For the first few minutes all went well, and the motor seemed to be working satisfactorily. The air ship answered the helm readily, and admiring exclamations rose from the crowd.... But as the vessel rose higher she was seen to fall off from the wind, while the aeronauts could be seen vainly endeavouring to keep her head on. Then M. Severo commenced throwing out ballast.... All this time the ship was gradually soaring higher and higher until, just as it was over the Montparnasse Cemetery, at the height of 2,000 feet, a sheet of flame was seen to shoot up from one of the motors, and instantly the immense silk envelope containing 9,000 cubic feet of hydrogen was enveloped in leaping tongues of fire.... As soon as the flames came in contact with the gas a tremendous explosion followed, and in an instant all that was left of the air ship fell to the earth." Both aeronauts were dashed to pieces. It was thought that the fatality was caused through faulty construction, the escape valve for the gas being situated only about nine feet from the motor. It was announced by Count de la Vaulx that during the summer of 1901 he would attempt to cross the Mediterranean by a balloon, provisioned for three weeks, maintaining communication with the coast during his voyage by wireless telegraphy and other methods of signalling. He was to make use of the "Herve Deviator," or steering apparatus, which may be described as a series of cup-shaped plates dipping in the water at the end of a trail rope. By means of controlling cords worked from the car, the whole series of plates could be turned at an angle to the direction of the wind, by which the balloon's course would be altered. Count de la Vaulx attempted this grand journey on October 12th, starting from Toulon with the intention of reaching Algiers, taking the precauti-on, however, of having a cruiser in attendance. When fifty miles out from Marseilles a passing steamer received from the balloon the signal, "All's well"; but the wind had veered round to the east, and, remaining persistently in this quarter, the Count abandoned his venture, and, signalling to the cruiser, succeeded in alighting on her deck, not, however, before he had completed the splendid and record voyage of 41 hours' durati-on.

CHAPTER XXVII. THE POSSIBILITIES OF BALLOONS IN WARFARE.

Clearly the time has not yet arrived when the flying machine will be serviceable in war. Yet we are not without those theorisers who, at the present moment, would seriously propose schemes for conveying dynamite and other explosives by air ship, or dropping them over hostile forces or fortresses, or even fleets at sea. They go yet further, and gravely discuss the point whether such warfare would be legitimate. We, however, may say at once, emphatically, that any such scheme is simply impracticable. It must be abundantly evident that, so far, no form of dirigible air ship exists which could be relied on to carry out any required manoeuvre in such atmospheric conditions as generally prevail. If, even in calm and favourable weather, more often than not motors break down, or gear carries away, what hope is there for any aerial craft which would attempt to battle with such wind currents as commonly blow aloft?

And when we turn to the balloon proper, are chances greatly improved? The eminently practical aeronaut, John Wise, as was told in Chapter XII., prepared a scheme for the reduction of Vera Cruz by the agency of a balloon. Let us glance at it. A single balloon was to suffice, measuring 100 feet in diameter, and capable of raising in the gross 30,000 lbs. To manoeuvre this monstrous engine he calculates he would require a cable five miles long, by means of which he hoped, in some manner, to work his way directly over the fortress, and to remain poised at that point at the height of a mile in the sky. Once granted that he could arrive and maintain himself at that position, the throwing out of combustibles would be simple, though even then the spot where they would alight after the drop of a mile would be by no means certain. It is also obvious that a vast amount of gas would have to be sacrificed to compensate for the prodigal discharge of ballast in the form of missiles.

The idea of manoeuvring a balloon in a wind, and poising it in the manner suggested, is, of course, preposterous; and when one considers the attempt to aim bombs from a moving balloon high in air the case becomes yet more absurd. Any such missile would partake of the motion of the balloon itself, and it would be impossible to tell where it would strike the earth.

To give an example which is often enough tried in balloon travel when the ground below is clear. A glass bottle (presumably empty) is cast overboard and its fall watched. It is seen not to be left behind, but to keep pace with the balloon, shrinking gradually to an object too small to be discerned, except when every now and then a ray of sunlight reflected off it reveals it for a moment as it continues to plunge downwards. After a very few seconds the impression is that it is about to reach the earth, and the eye forms a guess at some spot which it will strike; but the spot is quickly passed, and the bottle travels far beyond across a field, over the further fence, and vastly further yet; indeed, inasmuch as to fall a mile in air a heavy body may take over twenty seconds—and twenty seconds is long to those who watch—it is often impossible to tell to two or three fields where it will finally settle.

All this while the risk that a balloon would run of being riddled by bullets, shrapnel, or pom-poms has not been taken into account, and as to the estimate of this risk there is some difference of opinion. The balloon corps and the artillery apparently approach the question with different bias. On the one hand, it is stated with perfect truth that a free balloon, which is generally either rising or falling, as well as moving across country, is a hard object to hit, and a marksman would only strike it with a chance or blundering shot; but, on the other hand let us take the following report of three years ago.

The German artillery had been testing the efficiency of a quick-firing gun when used against a balloon, and they decided that the latter would have no chance of escape except at night. A German kite-balloon was kept moving at an altitude of 600 metres, and the guns trained upon it were distant 3,000 metres. It was then stated that after the third discharge of the rapid firing battery the range was found, when all was at once over with the balloon; for, not only was it hit with every discharge, but it was presently set on fire and annihilated.

But, in any case, the antique mode of keeping a balloon moored at any spot as a post of observation must be abandoned in modern warfare. Major Baden-Powell, speaking from personal experience in South Africa, has shown how dangerous, or else how useless, such a form of reconnaissance has become. "I remember," he says, "at the battle of Magersfontein my company was lying down in extended order towards the left of our line. We were perfectly safe from musketry fire, as we lay, perhaps, two miles from the Boer trenches, which were being shelled by some of our guns close by. The enemy's artillery was practically silent. Presently, on looking round, I descried our balloon away out behind us about two miles off. Then she steadily rose and made several trips to a good height, but what could be seen from that distance? When a large number of our troops were ranged up within 800 yards of the trenches, and many more at all points behind them, what useful information could be obtained by means of the balloon four miles off?"

The same eminent authority insists on the necessity of an observing war balloon making short ascents. The balloon, in his opinion, should be allowed to ascend rapidly to its full height, and with as little delay as possible be hauled down again. Under these conditions it may then be well worth testing whether the primitive form of balloon, the Montgolfier, might not be the most valuable. Instead of being made, as the war balloon is now, of fragile material, and filled with costly gas difficult to procure, and which has to be conveyed in heavy and cumbersome cylinders, a hot air balloon could be rapidly carried by hand anywhere where a few men could push their way. It is of strong material, readily mended if torn, and could be inflated for short ascents, if not by mere brush wood, then by a portable blast furnace and petroleum.

But there is a further use for balloons in warfare not yet exploited. The Siege of Paris showed the utility of free balloons, and occasions arise when their use might be still further extended. The writer pointed out that it might have been very possible for an aeronaut of experience, by choosing the right weather and the right position along the British lines, to have skilfully manoeuvred a free balloon by means of upper currents, so as to convey all-important intelligence to besieged Mafeking, and he proved that it would have sufficed if the balloon could have been "tacked" across the sky to within some fifteen miles of the desired goal.

The mode of signalling which he proposed was by means of a "collapsing drum," an instrument of occasional use in the Navy. A modification of this instrument, as employed by the writer, consisted of a light, spherical, drum-shaped frame of large size, which, when covered with dark material and hung in the clear below the car of a lofty balloon, could be well seen either against blue sky or grey at a great distance. The so-called drum could, by a very simple contrivance, readily worked from the car, be made to collapse into a very inconspicuous object, and thus be capable of displaying Morse Code signals. A long pause with the drum extended—like the long wave of a signalling flag—would denote a "dash," and a short pause a "dot," and these motions would be at once intelligible to anyone acquainted with the now universal Morse Code system.

Provided with an apparatus of the kind, the writer made an ascent from Newbury at a time when the military camps were lying on Salisbury Plain at a distance of nearly twenty miles to the south-west. The ground wind up to 2,500 feet on starting was nearly due north, and would have defeated the attempt; again, the air stream blowing above that height was nearly due east, which again would have proved unsuitable. But it was manifestly possible to utilise the two currents, and with good luck to zig-zag one's course so as to come within easy signalling distance of the various camps; and, as a matter of fact, we actually passed immediately over Bulford Camp, with which we exchanged signals, while two other camps lay close to right and left of us. Fortune favouring us, we had actually hit our mark, though it would have been sufficient for the experiment had our course lain within ten miles right or left.

Yet a further use for the balloon in warfare remains untried in this country. Acting under the advice of experts in the Service, the writer, in the early part of the present year, suggested to the Admiralty the desirability of experimenting with balloons as a means of detecting submarine engines of war. It is well known that reefs and shoals can generally be seen from a cliff or mast head far more clearly than from the deck or other position near the surface of the water. Would not, then, a balloon, if skilfully manoeuvred, serve as a valuable post of observation? The Admiralty, in acknowledging the communication, promised to give the matter their attention; but by the month of June the Press had announcements of how the self-same experiments had been successfully carried through by French authorities, while a few days later the Admiralty wrote, "For the present no need is seen for the use of a captive balloon to detect submarines."

Among many and varied ballooning incidents which have occurred to the writer, there are some which may not unprofitably be compared with certain experiences already recorded of other aeronauts. Thunderstorms, as witnessed from a balloon, have already been casually described, and it may reasonably be hoped that the observations which have, under varying circumstances, been made at high altitudes may throw some additional light on this familiar, though somewhat perplexing, phenomenon.

To begin with, it seems a moot point whether a balloon caught in a thunderstorm is, or is not, in any special danger of being struck. It has been argued that immunity under such circumstances must depend upon whether a sufficiently long time has elapsed since the balloon left the earth to allow of its becoming positively electrified by induction from the clouds or by rain falling upon its surface. But there are many other points to be considered. There is the constant escape of gas from the mouth; there is the mass of pointed metal in the anchor; and, again, it is conceivable that a balloon rapidly descending out of a thunderstorm might carry with it a charge residing on its moistened surface which might manifest itself disastrously as the balloon reached the earth.

Instances seem to have been not infrequent of balloons encountering thunderstorms; but, unfortunately, in most cases the observers have not had any scientific training, or the accounts which are to hand are those of the type of journalist who is chiefly in quest of sensational copy.

Thus there is an account from America of a Professor King who made an ascent from Burlington, Iowa, just as a thunderstorm was approaching, with the result that, instead of scudding away with the wind before the storm, he was actually, as if by some attraction, drawn into it. On this his aim was to pierce through the cloud above, and then follows a description which it is hard to realise:—"There came down in front of him, and apparently not more than 50 feet distant, a grand discharge of electricity." Then he feels the car lifted, the gas suddenly expands to overflowing, and the balloon is hurled through the cloud with inconceivable velocity, this happening several times, with tremendous oscillations of the car, until the balloon is borne to earth in a torrent of rain. We fancy that many practical balloonists will hardly endorse this description.

But we have another, relating to one of the most distinguished aeronauts, M. Eugene Godard, who, in an ascent with local journalists, was caught in a thunderstorm. Here we are told—presumably by the journalists—that "twice the lightning flashed within a few yards of the terror-stricken crew."

Once again, in an ascent at Derby, a spectator writes:—"The lightning played upon the sphere of the balloon, lighting it up and making things visible through it." This, however, one must suppose, can hardly apply to the balloon when liberated.

But a graphic description of a very different character given in the "Quarterly Journal of the Royal Meteorological Society" for January, 1901, is of real value. It appears that three lieutenants of the Prussian Balloon Corps took charge of a balloon that ascended at Berlin, and, when at a height of 2,300 feet, became enveloped in the mist, through which only occasional glimpses of earth were seen. At this point a sharp, crackling sound was heard at the ring, like the sparking of a huge electrical machine, and, looking up, the voyagers beheld sparks apparently some half-inch thick, and over two feet in length, playing from the ring. Thunder was heard, but—and this may have significance—only before and after the above phenomenon.

Another instructive experience is recorded of the younger Green in an ascent which he made from Frankfort-on-the-Maine. On this occasion he relates that he encountered a thunderstorm, and at a height of 4,400 feet found himself at the level where the storm clouds were discharging themselves in a deluge. He seems to have had no difficulty in ascending through the storm into the clear sky above, where a breeze from another quarter quickly carried him away from the storm centre.

This co-existence, or conflict of opposite currents, is held to be the common characteristic, if not the main cause, of thunderstorms, and tallies with the following personal experience. It was in typical July weather of 1900 that the writer and his son, accompanied by Admiral Sir Edmund Fremantle and Mr. Percival Spencer, made an evening ascent from Newbury. It had been a day

Many other aerial vessels might be mentioned. Mr. T. H. Bastin, of Clapham, has been engaged for many years on a machine which should imitate bird flight as nearly as this may be practicable.

Baron Bradsky aims at a navigable balloon on an ambitious scale. M. Tatin is another candidate for the Deutsch prize. Of Dr. Barton's air ship more is looked for, as being designed for the War Office. It is understood that the official requirements demand a machine which, while capable of transporting a man through the air at a speed of 13 miles an hour, can remain fully inflated for 48 hours. One of the most sanguine, as well as enterprising, imitators of Santos Dumont was a fellow countryman, Auguste Severo. Of his machine during construction little could be gathered, and still less seen, from the fact that the various parts were being manufactured at different workshops, but it was known to be of large size and to be fitted with powerful motors. This was an ill-fated vessel. At an early hour on May 12th of this year, 1902, all Paris was startled by a report that M. Severo and his assistant, M. Sachet had been killed while making a trial excursion. It appears that at daybreak it had been decided that the favourable moment for trial had arrived. The machinery was got ready, and with little delay the air vessel was dismissed and rose quietly and of storms, but about 5 p.m., after what appeared to be a clearing shower, the sky brightened, and we sailed up into a cloudless heaven. The wind, at 3,000 feet, was travelling at some thirty miles an hour, and ere the distance of ten miles had been covered a formidable thunder pack was seen approaching and coming up dead against the wind. Nothing could be more evident than that the balloon was travelling rapidly with a lower wind, while the storm was being borne equally rapidly on an upper and diametrically opposite current. It proved one of the most severe thunderstorms remembered in the country. It brooded for five hours over Devizes, a few miles ahead. A homestead on our right was struck and burned to the ground, while on our left two soldiers were killed on Salisbury Plain. The sky immediately overhead was, of course, hidden by the large globe of the balloon, but around and beneath us the storm seemed to gather in a blue grey mist, which quickly broadened and deepened till, almost before we could realise it, we found ourselves in the very heart of the storm, the lightning playing all around us, and the sharp hail stinging our faces.

The countrymen below described the balloon as apparently enveloped by the lightning, but with ourselves, though the flashes were incessant, and on all sides, the reverberations of the thunder were not remarkable, being rather brief explosions in which they resembled the thunder claps not infrequently described by travellers on mountain heights.

The balloon was now descending from a double cause: the weight of moisture suddenly accumulated on its surface, and the very obvious down-rush of cold air that accompanied the storm of pelting hail. With a very limited store of ballast, it seemed impossible to make a further ascent, nor was this desirable. The signalling experiments on which we were intent could not be carried on in such weather. The only course was to descend, and though this was not at once practicable, o-wing to Savernake Forest being beneath us, we effected a safe landing in the first available clea-ring.

As has been mentioned, Mr. Glaisher and other observers have recorded several remarkable instances of opposite wind currents being met with at moderate altitudes. None, however, can have been more noteworthy or surprising than the following experience Of the writer on Whit Monday of 1899. The ascent was under an overcast sky, from the Crystal Palace at 3 p.m., at which hour a cold drizzle was settling in with a moderate breeze from the east. Thus, starting from the usual filling ground near the north tower, the balloon sailed over the body of the Palace, and thence over the suburbs towards the west till lost in the mist. We then ascended through 1,500 feet of dense, wetting cloud, and, emerging in bright sunshine, continued to drift for two hours at an average altitude of some 3,000 feet; 1,000 feet below us was the ill-defined, ever changing upper surface of the dense cloud floor, and it was no longer possible to determine our course, which we therefore assumed to have remained unchanged. At length, however, as a measure of prudence, we determined to descend through the clouds sufficiently to learn something of our whereabouts, which we reasonably expected to be somewhere in Surrey or Berks. On emerging, however, below the cloud, the first object that loomed out of the mist immediately below us was a cargo vessel, in the rigging of which our trail rope was entangling itself. Only by degrees the fact dawned upon us that we were in the estuary of the Thames, and beating up towards London once again with an cast wind. Thus it became evident that at the higher level, unknown to ourselves, we had been headed back on our course, for two hours, by a wind diametrically opposed to that blowing on the ground.

Two recent developments of the hot-air war balloon suggest great possibilities in the near future. One takes the form of a small captive, carrying aloft a photographic camera directed and operated electrically from the ground. The other is a self-contained passenger balloon of large dimensions, carrying in complete safety a special petroleum burner of great power. These new and important departures are mainly due to the mechanical genius of Mr. J. N. Maskelyne, who has patented and perfected them in conjunction with the writer.

CHAPTER XXVIII. THE CONSTITUTION OF THE AIR.

Some fair idea of the conditions prevailing in the upper air may have been gathered from the many and various observations already recorded. Stating the case broadly, we may assert that the same atmospheric changes with which we are familiar at the level of the earth are to be found also at all accessible heights, equally extensive and equally sudden.

Standing on an open heath on a gusty day, we may often note the rhythmic buffeting of the wind, resembling the assault of rolling billows of air. The evidence of these billows has been actually traced far aloft in balloon travel, when aeronauts, looking down on a wind-swept surface of cloud, have observed this surface to be thrown into a series of rolls of vapour, which were but vast and veritable waves of air. The interval between successive crests of these waves has on one occasion been estimated at approximately half a mile. We have seen how these air streams sometimes hold wide and independent sway at different levels. We have seen, too, how they sometimes meet and mingle, not infrequently attended with electrical disturbance

Through broad drifts of air minor air streams would seem often literally to "thread" their way, breaking up into filaments or wandering rills of air. In the voyage across Salisbury Plain lately described, while the balloon was being carried with the more sluggish current, a number of small parachutes were dropped out at frequent intervals and carefully watched. These would commonly attend the balloon for a little while, until, getting into some minor air stream, they would suddenly and rapidly diverge at such wide angles as to suggest that crossing our actual course there were side paths, down which the smaller bodies became wafted.

On another occasion the writer met with strongly marked and altogether exceptional evidence of the vehemence and persistence of these minor aerial streamlets. It was on an occasion in April weather, when a heavy overcast sky blotted out the upper heavens. In the cloud levels the wind was somewhat sluggish, and for an hour we travelled at an average speed of a little over twenty miles an hour, never higher than 3,000 feet. At this point, while flying over Hertfordshire, we threw out sufficient ballast to cause the balloon to rise clear of the hazy lower air, and coming under the full influence of the sun, then in the meridian, we shot upwards at considerable speed, and soon attained an altitude of three miles. But for a considerable portion of this climb—while, in fact, we were ascending through little less than a mile of our upward course—we were assailed by impetuous cross currents, which whistled through car and rigging and smote us fairly on the cheek. It was altogether a novel experience, and the more remarkable from the fact that our main onward course was not appreciably diverted.

Then we got above these currents, and remained at our maximum level, while we floated, still at only a moderate speed, the length of a county. The descent then began, and once again, while we dropped through the same disturbed region, the same far-reaching and obtrusive cross-current assailed us. It was quite obvious that the vehement currents were too slender to tell largely upon the huge surface of the balloon, as it was being swept steadily onwards by the main wind, which never varied in direction from ground levels up to the greatest height attained.

This experience is but confirmation of the story of the wind told by the wind gauges on the Forth Bridge. Here the maximum pressure measured on the large gauge of 300 square feet is commonly considerably less than that on the smaller gauge, suggesting that the latter must be due to threads of air of limited area and high velocity.

Further and very valuable light is thrown on the peculiar ways of the wind, now being considered, by Professor Langley in the special researches of his to which reference has already been made. This eminent observer and mathematician, suspecting that the old-fashioned instruments, which only told what the wind had been doing every hour, or at best every minute, gave but a most imperfect record, constructed delicate gauges, which would respond to every impulse and give readings from second to second.

In this way he established the fact that the wind, far from being a body of even approximate uniformity, is under most ordinary conditions irregular almost beyond conception. Further, that the greater the speed the greater the fluctuations, so that a high wind has to be regarded as "air moving in a tumultuous mass," the velocity at one moment perhaps forty miles an hour, then diminishing to an almost instantaneous calm, and then resuming. "In fact, in the very nature of the case, wind is not the result of one simple cause, but of an infinite number of impulses and changes, perhaps long passed, which are preserved in it, and which die only slowly away."

When we come to take observations of temperature we find the conditions in the atmosphere above us to be at first sight not a little complex, and altogether different in day and night hours. From observations already recorded in this volume—notably those of Gay Lussac, Welsh, and Glaisher—it has been made to appear that, in ascending into the sky in daytime, the temperature usually falls according to a general law; but there are found regions where the fall of temperature becomes arrested, such regions being commonly, though by no means invariably, associated with visible cloud. It is probable, however, that it would be more correct not to interpret the presence of cloud as causing manifestation of cold, but rather to regard the meeting of warm and cold currents as the cause of cloud.

The writer has experimented in the upper regions with a special form of air thermometer of great sensibility, designed to respond rapidly to slight variations of temperature. Testing this instrument on one occasion in a room of equable warmth, and without draughts, he was puzzled by seeing the index in a capillary tube suddenly mounting rapidly, due to some cause which was not apparent, till it was noticed that the parlour cat, attracted by the proceedings, had approached near the apparatus. The behaviour of this instrument when slung in the clear some distance over the side of the balloon car, and carefully watched, suggests by its fitful, sudden, and rapid changes that warmer currents are often making their way in such slender wandering rills as have been already pictured as permeating the broader air streams. During night hours conditions are reversed. The warmer air radiated off the earth through the day has then ascended. It will be found at different heights, lying in pools or strata, possibly resembling in form, could they be seen, masses of visible cloud.

The writer has gathered from night voyages instructive and suggestive facts with reference to the ascent of air streams, due to differences of temperature, particularly over London and the suburbs, and it is conceivable that in such ascending streams may lie a means of dealing successfully with visitations of smoke and fog.

One lesson taught by balloon travel has been that fog or haze will come or go in obedience to temperature variations at low levels. Thus thick haze has lain over London, more particularly over the lower parts, at sundown. Then through night hours, as the temperature of the lower air has become equalised, the haze has completely disappeared, but only to reassert itself at dawn.

A description of the very impressive experience of a night sail over London has been reserved, but should not be altogether omitted. Glaisher, writing of the spectacle as he observed it nearly forty years ago, describes London seen at night from a balloon at a distance as resembling a vast conflagration. When actually over the town, a main thoroughfare like the Commercial Road shone up like a line of brilliant fire; but, travelling westward, Oxford Street presented an appearance which puzzled him. "Here the two thickly studded rows of brilliant lights were seen on either side of the street, with a narrow, dark space between, and this dark space was bounded, as it were, on both sides by a bright fringe like frosted silver." Presently he discovered that this rich effect was caused by the bright illumination of the shop lights on the pavements.

London, as seen from a balloon on a clear moonlight night in August a year ago (1901), wore a somewhat altered appearance. There were the fairy lamps tracing out the streets, which, though dark centred, wore their silver lining; but in irregular patches a whiter light from electric arc lamps broadened and brightened and shone out like some pyrotechnic display above the black housetops. Through the vast town ran a blank, black channel, the river, winding on into distance, crossed here and there by bridges showing as bright bands, and with bright spots occasionally to mark where lay the river craft. But what was most striking was the silence. Though the noise of London traffic as heard from a balloon has diminished of late years owing to the better paving, yet in day hours the roar of the streets is heard up to a great height as a hard, harsh, grinding din. But at night, after the last 'bus has ceased to ply, and before the market carts begin lumbering in, the balloonist, as he sails over the town, might imagine that he was traversing a City of the Dead.

It is at such times that a shout through a speaking trumpet has a most startling effect, and more particularly a blast on a horn. In this case after an interval of some seconds a wild note will be flung back from the house-tops below, answered and re-answered on all sides as it echoes from roof to roof—a wild, weird uproar that awakes suddenly, and then dies out slowly far away.

Experiments with echoes from a balloon have proved instructive. If, when riding at a height, say, of 2,000 feet, a charge of gun-cotton be fired electrically 100 feet below the car, the report, though really as loud as a cannon, sounds no more than a mere pistol shot, possibly partly owing to the greater rarity of the air, but chiefly because the sound, having no background to reflect it, simply spends itself in the air. Then, always and under all conditions of atmosphere soever, there ensues absolute silence until the time for the echo back from earth has fully elapsed, when a deafening outburst of thunder rises from below, rolling on often for more than half a minute. Two noteworthy facts, at least, the writer has established from a very large number of trials: first, that the theory of aerial echoes thrown back from empty space, which physicists have held to exist constantly, and to be part of the cause of thunder, will have to be abandoned; and, secondly, that from some cause yet to be fully explained the echo back from the earth is always behind its time.

But balloons have revealed further suggestive facts with regard to sound, and more particularly with regard to the varying acoustic properties of the air. It is a familiar experience how distant sounds will come and go, rising and falling, often being wafted over extraordinary distances, and again failing altogether, or sometimes being lost at near range, but appearing in strength further away. A free balloon, moving in the profound silence of the upper air, becomes an admirable sound observatory. It may be clearly detected that in certain conditions of atmosphere, at least, there are what may be conceived to be aerial sound channels, through which sounds are momentarily conveyed with abnormal intensity. This phenomenon does but serve to give an intelligible presentment of the unseen conditions existing in the realm of air.

It would be reasonable to suppose that were an eye so constituted as to be able to see, say, cumulus masses of warmer air, strata mottled with traces of other gases, and beds of invisible matter in suspension, one might suppose that what we deem the clearest sky would then appear flecked with forms as many and various as the clouds that adorn our summer heavens.

But there is matter in suspension in the atmosphere which is very far from invisible, and which in the case of large towns is very commonly lying in thick strata overhead, stopping back the sunlight, and forming the nucleus round which noisome fogs may form. Experimenting with suitable apparatus, the writer has found on a still afternoon in May, at 2,000 feet above Kingston in Surrey, that the air was charged far more heavily with dust than that of the London streets the next day; and, again, at half a mile above the city in the month of August last dust, much of it being of a gross and even fibrous nature, was far more abundant than on grass enclosures in the town during the forenoon of the day following.

An attempt has been made to include England in a series of international balloon ascents arranged expressly for the purpose of taking simultaneous observations at a large number of stations over Europe, by which means it is hoped that much fresh knowledge will be forthcoming with respect to the constitution of the atmosphere up to the highest levels accessible by balloons manned and unmanned. It is very much to be regretted that in the case of England the attempt here spoken of has rested entirely on private enterprise. First and foremost in personal liberality and the work of organisation must be mentioned Mr. P. Y. Alexander, whose zeal in the progress of aeronautics is second to none in this country. Twice through his efforts England has been represented in the important work for which Continental nations have no difficulty in obtaining public grants. The first occasion was on November 8th, 1900, when the writer was privileged to occupy a seat in the balloon furnished by Mr. Alexander, and equipped with the most modern type of instruments. It was a stormy and fast voyage from the Crystal Palace to Halstead, in Essex, 48 miles in 40 minutes. Simultaneously with this, Mr. Alexander dismissed an unmanned balloon from Bath, which ascended 8,000 feet, and landed at Cricklade. Other balloons which took part in the combined experiment were two from Paris, three from Chalais Meudon, three from Strasburg, two from Vienna, two from Berlin, and two from St. Petersburg.

The section of our countrymen specially interested in aeronautics — a growing community — is represented by the Aeronautical Society, formed in 1865, with the Duke of Argyll for president, and for thirty years under the most energetic management of Mr. F. W. Brearey, succeeding whom as hon. secs. have been Major Baden-Powell and Mr. Eric S. Bruce. Mr. Brearey was one of the most successful inventors of flying models. Mr. Chanute, speaking as President of the American Society of Civil Engineers, paid him a high and well-deserved compliment in saying that it was through his influence that aerial navigation had been cleared of much rubbish and placed upon a scientific and firm basis.

Another community devoting itself to the pursuit of balloon trips and matters aeronautical generally is the newly-formed Aero Club, of whom one of the most prominent and energetic members is the Hon. C. S. Rolls.

It had been announced that M. Santos-Dumont would bring an air ship to England, and during the summer of the present year would give exhibitions of its capability. It was even rumoured that he might circle round St. Paul's and accomplish other aerial feats unknown in England. The promise was fulfilled so far as bringing the air ship to England was concerned, for one of his vessels which had seen service was deposited at the Crystal Palace. In some mysterious manner, however, never sufficiently made clear to the public, this machine was one morning found damaged, and M. Santos-Dumont has withdrawn from his proposed engagements.

In thus doing he left the field open to one of our own countrymen, who, in his first attempt at flight with an air ship of his own invention and construction, has proved himself no unworthy rival of the wealthy young Brazilian.

Mr. Stanley Spencer, in a very brief space of time, designed and built completely in the workshops of the firm an elongated motor balloon, 75 feet long by 20 feet diameter, worked by a screw and petrol motor. This motor is placed in the prow, 25 feet away from, and in front of, the safety valve, by which precaution any danger of igniting the escaping gas is avoided. Should, however, a collapse of the machine arise from any cause, there is an arrangement for throwing the balloon into the form of a parachute. Further, there is provided means for admitting air at will into the balloon, by which the necessity for much ballast is obviated.

Mr. Spencer having filled the balloon with pure hydrogen, made his first trial with this machine late in an evening at the end of June. The performance of the vessel is thus described in the Westminster Gazette:—"The huge balloon filled slowly, so that the light was rapidly failing when at last the doors of the big shed slid open and the ship was brought carefully out, her motor started, and her maiden voyage commenced. With Mr. Stanley Spencer in the car, she sailed gracefully down the football field, wheeled round in a circle—a small circle, too—and for perhaps a quarter of an hour sailed a tortuous course over the heads of a small but enthusiastic crowd of spectators. The ship was handicapped to some extent by the fact that in their anxiety to make the trial the aeronauts had not waited to inflate it fully, but still it did its work well, answered its helm readily, showed no signs of rolling, and, in short, appeared to give entire satisfaction to everybody concerned—so much so, indeed, that Mr. Stanley Spencer informed the crowd after the ascent that he was quite ready to take up any challenge that

M. Santos Dumont might throw down." Within a few weeks of this his first success Mr. Spencer was able to prove to the world that he had only claimed for his machine what its powers fully justified. On a still September afternoon, ascending alone, he steered his aerial ship in an easy and graceful flight over London, from the Crystal Palace to Harrow.

CHAPTER XXIX. CONCLUSION.

The future development of aerostation is necessarily difficult to forecast. Having reviewed its history from its inception we have to allow that the balloon in itself, as an instrument of aerial locomotion, remains practically only where it was 120 years ago. Nor, in the nature of the case, is this to be wondered at. The wind, which alone guides the balloon, is beyond man's control, while, as a source of lifting power, a lighter and therefore more suitable gas than hydrogen is not to be found in nature.

It is, however, conceivable that a superior mode of inflation may yet be discovered. Now that the liquefaction of gases has become an accomplished fact, it seems almost theoretically possible that a balloonist may presently be able to provide himself with an unlimited reserve of potential energy so as to be fitted for travel of indefinite duration. Endowed with increased powers of this nature, the aeronaut could utilise a balloon for voyages of discovery over regions of the earth which bar man's progress by any other mode of travel. A future Andree, provided with a means of maintaining his gas supply for six weeks, need have no hesitation in laying his course towards the North Pole, being confident that the winds must ultimately waft him to some safe haven. He could, indeed, well afford, having reached the Pole, to descend and build his cairn, or even to stop a week, if he so desired, before continuing on his way.

But it may fairly be claimed for the balloon, even as it now is, that a great and important future is open to it as a means for exploring inaccessible country. It may, indeed, be urged that Andree's task was, in the very nature of the case, well nigh impracticable, and his unfortunate miscarriage will be used as argument against such a method of exploration. But it must always be remembered that in Andree's case the rigours of climate which he was compelled to face were the most serious of all obstacles to balloon travel. The extreme cold would not only cause constant shrinkage of the gas, but would entail the deposition of a weight of moisture, if not of snow, upon the surface of the balloon, which must greatly shorten its life.

It would be entirely otherwise if the country it were sought to explore were in lower latitudes, in Australia, or within the vast unknown belt of earth lying nearer the equator. The writer's scheme for exploring the wholly unknown regions of Arabia is already before the public. The fact, thought to be established by the most experienced aeronauts of old times, and already referred to in these pages, that at some height a strong west wind is to be found blowing with great constancy all round the globe, is in accordance with the view entertained by modern meteorologists. Such a wind, too, may be expected to be a fairly fast wind, the calculation being that, as a general rule, the velocity of currents increases from the ground at the rate of about three miles per hour for each thousand feet of height; thus the chance of a balloon drifting speedily across the breadth of Arabia is a strong one, and, regarded in this light, the distance to be traversed is certainly not excessive, being probably well within the lasting power of such a balloon as that employed by Andree. If, for the sake of gas supply, Aden were chosen for the starting ground, then 1,200 miles E.N.E. would carry the voyager to Muscat; 1,100 miles N.E. by E. would land him at Sohar; while some 800 miles would suffice to take him to the seaboard if his course lay N.E. It must also be borne in mind that the Arabian sun by day, and the heat radiated off the desert by night, would be all in favour of the buoyancy of the balloon.

But there are other persistent winds that, for purposes of exploration, would prove equally serviceable and sure. From time immemorial the dweller on the Nile has been led to regard his river in the light of a benignant deity. If he wished to travel down its course he had but to entrust his vessel to the stream, and this would carry him. If, again, he wished to retrace his course, he had but to raise a sail, and the prevalent wind, conquering the flood, would bear him against the stream. This constant north wind, following the Nile valley, and thence trending still southward towards Uganda, has been regarded as a means to hand well adapted for the exploration of important unsurveyed country by balloon. This scheme has been conceived and elaborated by Major B.F.S. Baden-Powell, and, so far, the only apparent obstacle in the way has proved the lack of necessary funds.

It will be urged, however, that for purposes of exploration some form of dirigible balloon is desirable, and we have already had proof that where it is not sought to combat winds strongly opposed to their course such air ships as Santos-Dumont or Messrs. Spencer have already constructed acquit themselves well; and it requires no stretch of imagination to conceive that before the present century is closed many great gaps in the map of the world will have been filled in by aerial survey.

But, leaving the balloon to its proper function, we turn to the flying machine properly so called with more sanguine hopes of seeing the real conquest of the air achieved. It was as it were but yesterday when the air ship, unhampered by huge globes of gas, and controlled by mechanical means alone, was first fairly tried, yet it is already considered by those best able to judge that its ultimate success is assured.

This success rests now solely in the hands of the mechanical engineer. He must, and surely can, build the ship of such strength that some essential part does not at the critical moment break down or carry away. He may have to improve his motive power, and here, again, we do not doubt his cunning. Motor engines, self-contained and burning liquid fuel, are yet in their infancy, and the extraordinary emulation now existing in their production puts it beyond doubt that every year will see rapid improvement in their efficiency.

We do not expect, nor do we desire, that the world may see the fulfilment of the poet's dream, "Argosies of magic sails" or "Airy navies grappling in the central blue." We would not befog our vision of the future with any wild imaginings, seeking, as some have done, to see in the electricity or other hidden power of heaven the means for its subjugation by man; but it is far from unreasonable to hope that but a little while shall pass, and we shall have more perfect and reliable knowledge of the tides and currents in the vast ocean of air, and when that day may have come then it may be claimed that the grand problem of aerial navigation will be already solved.